WITHDRAWN

WILLIAMS... ...NATIONAL... ...Y

Compact
Atlas of the World

SECOND EDITION

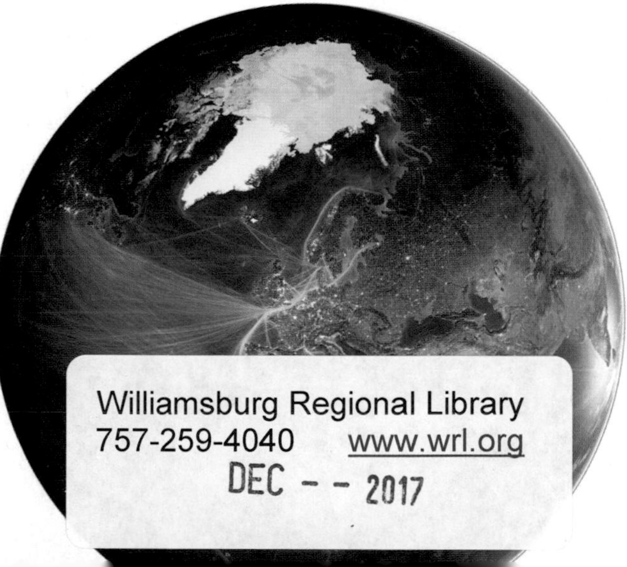

Williamsburg Regional Library
757-259-4040   www.wrl.org
DEC - - 2017

NATIONAL
GEOGRAPHIC

Compact

Atlas of the World

SECOND EDITION

NATIONAL GEOGRAPHIC
WASHINGTON, D.C.

CONTENTS

Continued on next page

AUSTRALIA & OCEANIA 146-161

POLAR REGIONS 162-167

WORLD FLAGS 168-175

APPENDIX 176-179

PLACE-NAME INDEX 180-253

ACKNOWLEDGMENTS 254-255

NORTH AMERICA
26–53

44

40–41

88–89

42–43

50–51

52–53

45

GREATER
OCEANIA
160–161

48–49

138–139

68–69

SOUTH AMERICA
54–73

70–71

72–73

110–111 Page(s) on
which map appears

KEY TO ATLAS MAPS

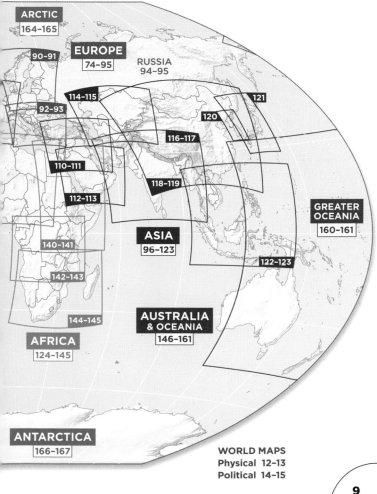

ARCTIC
164–165

EUROPE
74–95

90–91

RUSSIA
94–95

114–115

121

92–93

120

116–117

110–111

118–119

112–113

GREATER
OCEANIA
160–161

140–141

142–143

ASIA
96–123

144–145

AFRICA
124–145

AUSTRALIA
& OCEANIA
146–161

122–123

ANTARCTICA
166–167

WORLD MAPS
Physical 12–13
Political 14–15

NORTH AMERICA

EUROPE

VATICAN CITY———•

AFRICA

SOUTH AMERICA

•
└─ São Paulo

ANTARCTICA

WORLD

WORLD POLITICAL DATA

Total Number of Countries: 195

Largest Country by Area: Russia
17,098,242 sq km (6,601,665 sq mi)

Smallest Country by Area:
Vatican City   0.44 sq km (0.17 sq mi)

Most Populous Country: China
1,367,485,000

Least Populous Country: Vatican City
1,000

Largest Cities by Population:
Tokyo, Japan   38,001,000
Delhi, India   25,703,000
Shanghai, China   23,741,000
São Paulo, Brazil   21,066,000
Mumbai (Bombay) India   21,043,000

WORLD PHYSICAL DATA

Total Area: 510,072,000 sq km
(196,940,000 sq mi)

Land Area: 148,940,000 sq km
(57,506,000 sq mi), 29.1% of total

Water Area: 361,132,000 sq km
(139,434,000 sq mi), 70.9% of total

Equatorial Circumference: 40,075 km
(24,902 mi)

Polar Circumference: 40,008 km
(24,860 mi)

Equatorial Radius: 6,378 km (3,963 mi)

Polar Radius: 6,357 km (3,950 mi)

Highest Point: Mount Everest,
China-Nepal   8,850 m (29,035 ft)

Deepest Point: Challenger Deep,
Pacific Ocean   -10,984 m
(-36,037 ft)

Winkel Tripel Projection
Scale at the Equator

OCEAN

NANSEN BASIN

Norwegian Sea

Scandinavia

Barents Sea

Kara Sea

Laptev Sea

East Siberian Sea

SIBERIA

Yenisey

Ob

Lena

Ural Mountains

Irtysh

Bering Sea

Kamchatka Pen.

Sea of Okhotsk

Volga

Amur

EUROPE

Black Sea

Caspian Sea

Lake Baikal

Altay Mts.

Sakhalin

Hokkaido

Balkan Pen.

Anatolia (Asia Minor)

Tian Shan

GOBI

Korea

Honshu

ASIA

Yellow Sea

Kyushu

EMPEROR SEAMOUNTS

NORTH

Mediterranean Sea

Kunlun Mts.

HIMALAYA

Yangtze

East China Sea

Taiwan

PACIFIC

Mts.

Persian G.

Euphrates

INDIA

Indochina Peninsula

Philippine Islands

Philippine Sea

MID-PACIFIC MTS.

OCEAN

SAHARA

Arabian Peninsula

Arabian Sea

Bay of Bengal

Mariana Trench

AFRICA

Tigris

Caroline Islands

MICRONESIA

Red Sea

Ethiopian Highlands

Sri Lanka (Ceylon)

South China Sea

EQUATOR

Congo Basin

L. Victoria

Sumatra

Borneo

Indonesia

New Guinea

MELANESIA

L. Tanganyika

MID-INDIAN RIDGE

INDIAN

Java

Cape York Pen.

L. Malawi (Lake Nyasa)

Madagascar

NINETYEAST RIDGE

Coral Sea

Fiji Is.

Kalahari Desert

OCEAN

Western Plateau

AUSTRALIA

Great Dividing Range

Cape of Good Hope

SOUTHWEST INDIAN RIDGE

CROZET BASIN

Great Australian Bight

North I.

SOUTHEAST INDIAN RIDGE

Tasman Sea

New Zealand

KERGUELEN PLATEAU

SOUTH INDIAN BASIN

Tasmania

South I.

CAMPBELL PLATEAU

ENDERBY PLAIN

Queen Maud Land

Wilkes Land

ANTARCTICA

Transantarctic Mts.

30° 60° 90° 120° 150° 180° 90°

0		2000		4000 MILES

0		2000		4000 KILOMETERS

Global Land Cover Classes
Three characteristics underlie these categories: life-form (woody, herbaceous, or bare); leaf type (needle or broad); and leaf duration (evergreen or deciduous).

Evergreen needleleaf forest

Evergreen broadleaf forest

Deciduous needleleaf forest

Deciduous broadleaf forest

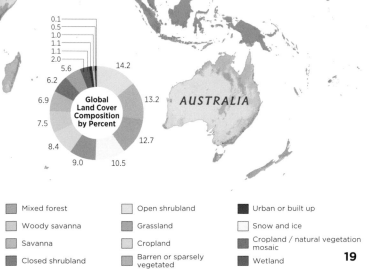

Satellite data provide the most reliable picture of global vegetative cover over time. Few natural communities of plants and animals have remained the same; most have been altered by humans. The "natural" vegetation reflects what would grow there, given ideal conditions. By recording the data at different wavelengths of the electromagnetic spectrum, scientists can derive land cover type through spectral variation. Changes in vegetation are captured, contributing to studies in conservation, biodiversity assessments, and land resource management.

ASIA

AUSTRALIA

Global Land Cover Composition by Percent

0.1
0.5
1.0
1.1
1.1
2.0
5.6
6.2
6.9
7.5
8.4
9.0
10.5
12.7
13.2
14.2

Mixed forest

Woody savanna

Savanna

Closed shrubland

Open shrubland

Grassland

Cropland

Barren or sparsely vegetated

Urban or built up

Snow and ice

Cropland / natural vegetation mosaic

Wetland

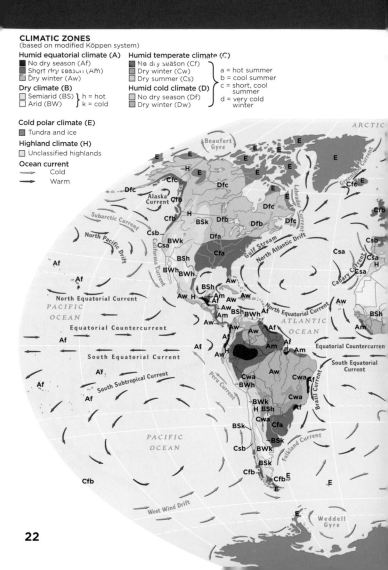

CLIMATIC ZONES
(based on modified Köppen system)

Humid equatorial climate (A)
- ■ No dry season (Af)
- ■ Short dry season (Am)
- □ Dry winter (Aw)

Dry climate (B)
- □ Semiarid (BS) } h = hot
- □ Arid (BW) } k = cold

Humid temperate climate (C)
- □ No dry season (Cf)
- □ Dry winter (Cw)
- □ Dry summer (Cs)

Humid cold climate (D)
- □ No dry season (Df)
- □ Dry winter (Dw)

a = hot summer
b = cool summer
c = short, cool summer
d = very cold winter

Cold polar climate (E)
- ■ Tundra and ice

Highland climate (H)
- □ Unclassified highlands

Ocean current
- → Cold
- → Warm

ARCTIC

Beaufort Gyre

Greenland Current

Alaska Current

Labrador Current

Subarctic Current

North Pacific Drift

Gulf Stream

North Atlantic Drift

California Current

Canary Current

North Equatorial Current

North Equatorial Current

PACIFIC OCEAN

Equatorial Countercurrent

ATLANTIC OCEAN

Equatorial Countercurrent

South Equatorial Current

South Equatorial Current

South Subtropical Current

Peru Current

Brazil Current

Falkland Current

PACIFIC OCEAN

West Wind Drift

Weddell Gyre

E E E E E E E E
Dfc Cfc Dfc Cfb Cfe
H Cfb H Dfb Dfc Dfc Cfb
Csb BSk Dfb Dfc Csb
Csa BWk Dfa Csa Csa
Af Cfa Csa H Csa
Af BSh Aw BSh H
BWh Aw H Am Aw
Af BWh Aw Am BSh BWh Af Aw BSh
Aw Af Am Am
Af Aw Af Af Am
Af Aw H Af Am
Af Aw Cwa
Af Cwa
BWh BWk Cwa
BSk H BSh Af
Cwa Cfa
Af BSk BSk
Csb BWk
BSk
Cfb Cfb E
E E

22

The Köppen system,
used here, classifies Earth's
climatic zones based on
precipitation, temperature, and
vegetation. These zones can
shift over time.

Climate is the average of the
elements of weather over
time. Climatic patterns are
established primarily by the
energy of the sun and the distri-
bution of solar radiation, which
is greatest at the Equator and
least at the Poles, and is modified
by altitude and distance from
the sea.

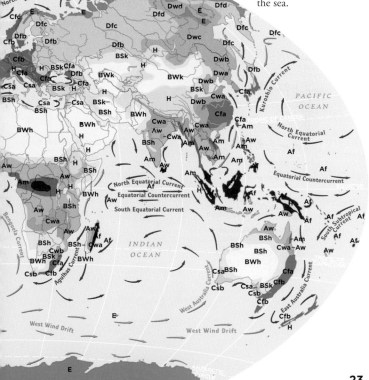

Hottest Recorded Temperature
Furnace Creek Ranch (Death Valley), California, U.S.: 56.7°C (134°F), October 7, 1913

Greatest Tidal Range ■
Bay of Fundy, Nova Scotia, Canada: 16 m (53 ft)

Most Crowded Country
Monaco: 18,800 per sq km (48,700 per sq mi)

Least Populous Country
Vatican City: 1,000 people

Smallest Country
Vatican City: 0.44 sq km (0.17 sq mi)

Largest Canyon
Grand Canyon, Arizona, U.S.: 446 km (277 mi) long along river, 180 m (600 ft) to 29 km (18 mi) wide, about 1.6 km (1 mi) deep

Highest Waterfall
Angel Falls, Venezuela: 979 m (3,212 ft)

Longest Mountain Range (submarine)
Mid-Ocean Ridge approximately 60,000 km (37,000 mi), encircles the Earth, mostly along the seafloor

Largest Drainage Basin
Amazon, South America: 6,145,186 sq km (2,372,670 sq mi)

Driest Place ■
Arica, Atacama Desert, Chile: rainfall barely measurable

Longest Mountain Range (continental)
Andes, South America: approximately 7,600 km (4,700 mi)

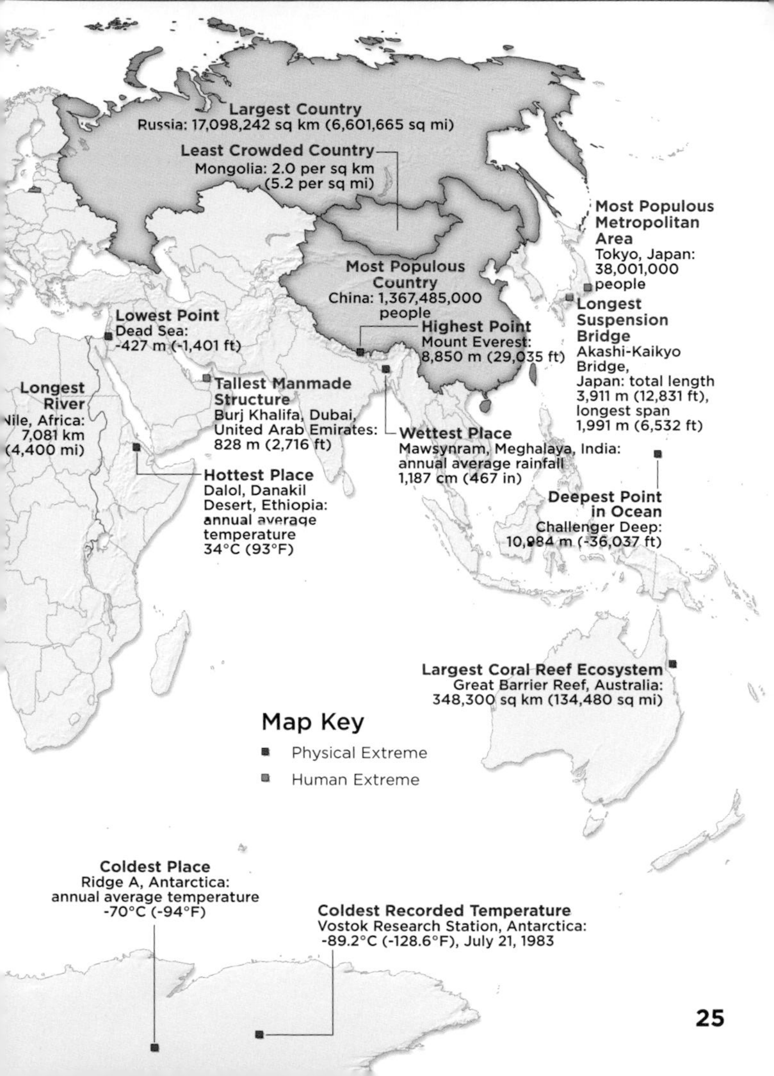

Largest Country
Russia: 17,098,242 sq km (6,601,665 sq mi)

Least Crowded Country
Mongolia: 2.0 per sq km
(5.2 per sq mi)

Most Populous Metropolitan Area
Tokyo, Japan: 38,001,000 people

Most Populous Country
China: 1,367,485,000 people

Highest Point
Mount Everest: 8,850 m (29,035 ft)

Longest Suspension Bridge
Akashi-Kaikyo Bridge, Japan: total length 3,911 m (12,831 ft), longest span 1,991 m (6,532 ft)

Lowest Point
Dead Sea: -427 m (-1,401 ft)

Longest River
Nile, Africa: 7,081 km (4,400 mi)

Tallest Manmade Structure
Burj Khalifa, Dubai, United Arab Emirates: 828 m (2,716 ft)

Wettest Place
Mawsynram, Meghalaya, India: annual average rainfall 1,187 cm (467 in)

Hottest Place
Dalol, Danakil Desert, Ethiopia: annual average temperature 34°C (93°F)

Deepest Point in Ocean
Challenger Deep: 10,984 m (-36,037 ft)

Largest Coral Reef Ecosystem
Great Barrier Reef, Australia: 348,300 sq km (134,480 sq mi)

Map Key

- Physical Extreme
- Human Extreme

Coldest Place
Ridge A, Antarctica: annual average temperature -70°C (-94°F)

Coldest Recorded Temperature
Vostok Research Station, Antarctica: -89.2°C (-128.6°F), July 21, 1983

Denali
(Mount McKinley)
6,190 m
(20,310 ft)

Yukon

Mackenzie

Peace

C A N A D A

Lake
Superior

Missouri

Mississippi

Death Valley
-86 m (-282 ft)

Los Angeles

U N I T E D   S T A T E S

Mexico
City

NORTH AMERICA

GEOGRAPHIC EXTREMES

CONTINENTAL POLITICAL FACTS

TOTAL NUMBER OF COUNTRIES: 23

LARGEST COUNTRY BY AREA: Canada
9,984,670 sq km (3,855,103 sq mi)

SMALLEST COUNTRY BY AREA: St. Kitts and
Nevis 261 sq km (101 sq mi)

MOST POPULOUS COUNTRY: United States
322,560,000

LEAST POPULOUS COUNTRY: St. Kitts and
Nevis 52,000

LARGEST URBAN AREAS BY POPULATION:
Mexico City, Mexico 20,999,000
New York, United States 18,593,000
Los Angeles, United States 12,310,000
Chicago, United States 8,745,000
Toronto, Canada 5,993,000

CONTINENTAL PHYSICAL FACTS

AREA: 24,474,000 sq km (9,449,000 sq mi)

HIGHEST POINT: Denali (Mount McKinley),
Alaska, United States 6,190 m
(20,310 ft)

LOWEST POINT: Death Valley, California,
United States -86 m (-282 ft)

LONGEST RIVERS:
Mississippi-Missouri 5,970 km
(3,710 mi)
Mackenzie-Peace 4,250 km (2,640 mi)
Yukon 3,220 km (2,000 mi)

LARGEST NATURAL LAKES:
Lake Superior 82,100 sq km
(31,700 sq mi)
Lake Huron 59,600 sq km
(23,000 sq mi)
Lake Michigan 57,800 sq km
(22,300 sq mi)

Lake Huron
Toronto
New York
Lake Michigan
Chicago

**ST. KITTS
AND NEVIS**

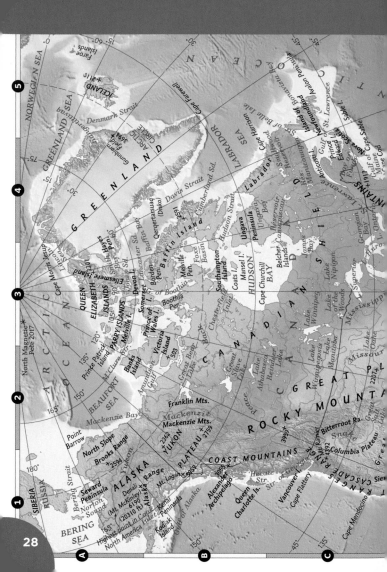

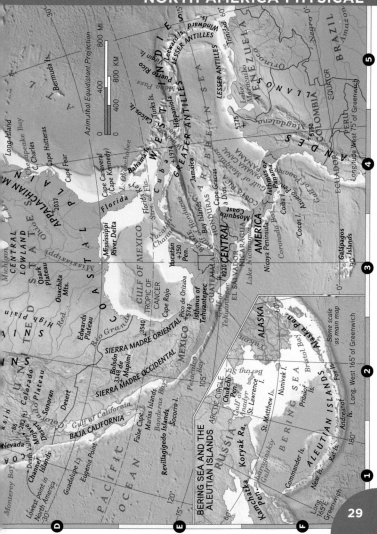

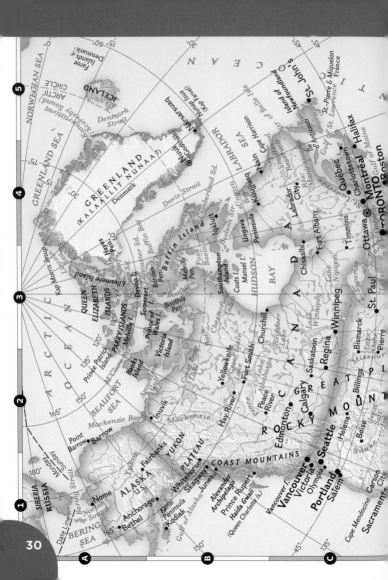

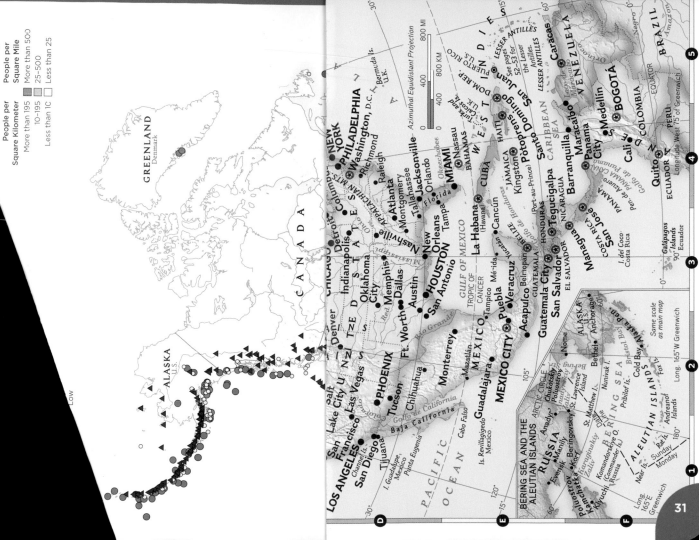

Population Density

People per Square Kilometer
- More than 195
- 10–195
- Less than 10

People per Square Mile
- More than 500
- 25–500
- Less than 25

GREENLAND
Denmark

CANADA

ALASKA
U.S.

Low

U N I T E D   S T A T E S

ROCKY MTS.
Columbia
Seattle

APPALACHIAN MTS.

CHICAGO
Detroit
Indianapolis
Oklahoma City
Memphis
Dallas
Ft. Worth
Austin
San Antonio
HOUSTON
New Orleans
Nashville
Atlanta
Montgomery
Tallahassee
Jacksonville
Orlando
Tampa
MIAMI

NEW YORK
PHILADELPHIA
Washington, D.C.
Richmond
Raleigh

Denver
Salt Lake City
Las Vegas
PHOENIX
Tucson
San Francisco
LOS ANGELES
San Diego

Ohio
Mississippi
Red River
Rio Grande

Florida
L. Okeechobee

Bermuda Is.
U.K.

Channel Is.
I. Guadalupe,
Mexico
Tijuana
Mexicali
Cabo Falso
Punta Eugenia
Baja California
Golfo de California
Is. Revillagigedo
Mexico

M E X I C O
MEXICO CITY
Guadalajara
Mazatlán
Chihuahua
Monterrey
Puebla
Veracruz
Tampico
Acapulco
Mérida
Cancún

GULF OF MEXICO

TROPIC OF CANCER

La Habana
(Havana)
CUBA
Nassau
BAHAMAS

JAMAICA
Kingston
HAITI
Port-au-Prince
DOM. REP.
Santo Domingo
PUERTO RICO
U.S.
San Juan

Turks and Caicos Is.
U.K.

W   E   S   T   I   N   D   I   E   S

LESSER ANTILLES
see pages 52–53 for the Lesser Antilles
LESSER ANTILLES

CARIBBEAN SEA

BELIZE
Belmopán
GUATEMALA
Guatemala City
EL SALVADOR
San Salvador
HONDURAS
Tegucigalpa
NICARAGUA
Managua
COSTA RICA
San José
PANAMA
Panama City
PANAMA CANAL

Golfo de Honduras

I. del Coco
Costa Rica

Galápagos Islands
Ecuador

Caracas
VENEZUELA
Maracaibo
Barranquilla
Medellín
BOGOTÁ
Cali
COLOMBIA
Quito
ECUADOR
PERU

BRAZIL
Amazon
Rio Negro

Orinoco

Golfo de Venezuela
Lago de Maracaibo
Golfo de Panamá

EQUATOR

Longitude West 75° of Greenwich

Azimuthal Equidistant Projection

800 MI
800 KM
400
0

BERING SEA AND THE ALEUTIAN ISLANDS

ARCTIC CIRCLE

RUSSIA
Chukotskiy
Anadyr'
Poluostrov
Provideniya

ALASKA
U.S.
Nome
Bethel
Anchorage
Cold Bay

BERING SEA

St. Lawrence Island
St. Matthew I.
Nunivak I.
Pribilof Is.

Fox Is.
ALEUTIAN ISLANDS
Andreanof Islands
Rat Is.
Near Is.

Bristol Bay
Alaska Pen.

Evensk
Magadan
Manily
Korf
Beringovskiy
Khatyrka
Kamchatka
Petropavlovsk-Kamchatskiy
Komandorskie (R.)
Commander Is.
Russia
Kamchatskiy Zaliv

Long. 165°W Greenwich
Long. 165°E Greenwich
Long. 180° Greenwich

Same scale as main map

Sunday
Monday

PACIFIC OCEAN

31

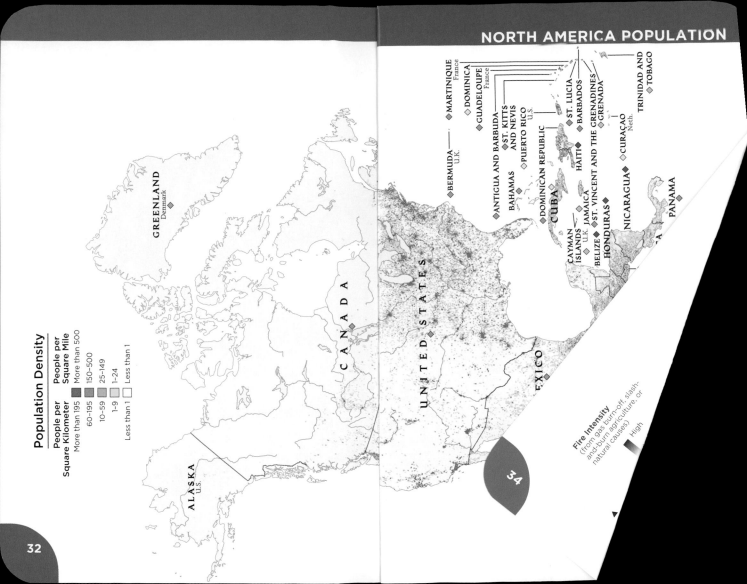

Population Density

People per Square Kilometer | **People per Square Mile**

More than 195 — More than 500
60-195 — 150-500
10-59 — 25-149
1-9 — 1-24
Less than 1 — Less than 1

GREENLAND
Denmark

ALASKA
U.S.

C A N A D A

U N I T E D   S T A T E S

MEXICO

BERMUDA
U.K.

MARTINIQUE
France

DOMINICA

GUADELOUPE
France

ANTIGUA AND BARBUDA

ST. KITTS
AND NEVIS

BAHAMAS

PUERTO RICO
U.S.

DOMINICAN REPUBLIC

CUBA

ST. LUCIA

BARBADOS

GRENADA

ST. VINCENT AND THE GRENADINES

CURAÇAO
Neth.

TRINIDAD AND
TOBAGO

HAITI

CAYMAN
ISLANDS

JAMAICA
U.K.

BELIZE

HONDURAS

NICARAGUA

PANAMA

Fire Intensity
(from gas burn-off, slash-and-burn agriculture, or natural causes)

High

32

34

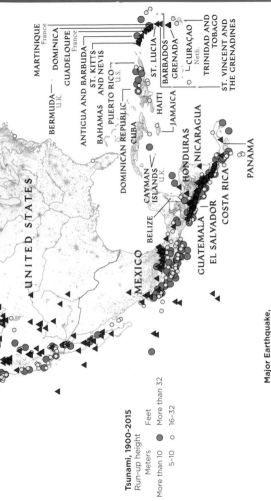

MARTINIQUE
France

DOMINICA

GUADELOUPE
France

BERMUDA
U.K.

ST. KITTS
AND NEVIS

ANTIGUA AND BARBUDA

BAHAMAS

PUERTO RICO
U.S.

ST. LUCIA

BARBADOS

GRENADA

CURAÇAO
Neth.

TRINIDAD AND
TOBAGO

ST. VINCENT AND
THE GRENADINES

HAITI

JAMAICA

DOMINICAN REPUBLIC

CUBA

CAYMAN
ISLANDS
U.K.

BELIZE

HONDURAS

NICARAGUA

PANAMA

GUATEMALA

EL SALVADOR

COSTA RICA

MEXICO

UNITED STATES

Tsunami, 1900-2015
Run-up height

Meters	Feet
More than 10 ●	More than 32
5-10 ○	16-32

**Major Earthquake,
1900-2015**
Moment magnitude

● More than 7.0

○ 6.0-7.0

Volcano

▲

Land Cover

- Evergreen needleleaf forest
- Evergreen broadleaf forest
- Deciduous needleleaf forest
- Deciduous broadleaf forest
- Mixed forest
- Woody savanna
- Savanna
- Closed shrubland
- Open shrubland
- Grassland
- Cropland
- Barren or sparsely vegetated
- Urban or built-up
- Snow and ice
- Cropland/natural vegetation mosaic
- Wetland
- ○ Urban area with more than 5 million inhabitants

ALASKA
U.S.

CANADA

GREENLAND
Denmark

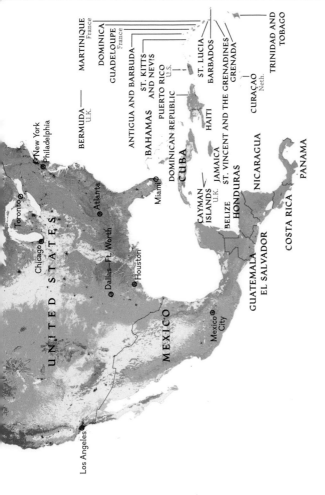

MARTINIQUE
France

DOMINICA

GUADELOUPE
France

BERMUDA
U.K.

ST. KITTS
AND NEVIS

ANTIGUA AND BARBUDA

PUERTO RICO
U.S.

BAHAMAS

DOMINICAN REPUBLIC

CUBA

ST. LUCIA

BARBADOS

ST. VINCENT AND THE GRENADINES
GRENADA

CURAÇAO
Neth.

TRINIDAD AND
TOBAGO

HAITI

JAMAICA

CAYMAN
ISLANDS
U.K.

BELIZE

HONDURAS

NICARAGUA

PANAMA

COSTA RICA

GUATEMALA
EL SALVADOR

MEXICO

New York
Philadelphia

Toronto

Atlanta

Chicago

Miami

Dallas–Ft. Worth

Houston

Mexico
City

Los Angeles

UNITED STATES

NORTH AMERICA CLIMATE

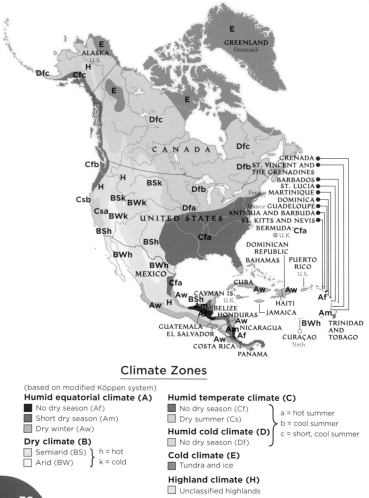

Climate Zones

(based on modified Köppen system)

Humid equatorial climate (A)
- ◼ No dry season (Af)
- ◼ Short dry season (Am)
- ◻ Dry winter (Aw)

Dry climate (B)
- ◻ Semiarid (BS) ⎱ h = hot
- ◻ Arid (BW) ⎰ k = cold

Humid temperate climate (C)
- ◼ No dry season (Cf)
- ◼ Dry summer (Cs)

Humid cold climate (D)
- ◻ No dry season (Df)

⎫ a = hot summer
⎬ b = cool summer
⎭ c = short, cool summer

Cold climate (E)
- ◼ Tundra and ice

Highland climate (H)
- ◻ Unclassified highlands

GREENLAND
Denmark

ALASKA
U.S.

C A N A D A

GRENADA
ST. VINCENT AND
THE GRENADINES
BARBADOS
ST. LUCIA
France MARTINIQUE
DOMINICA
France GUADELOUPE
ANTIGUA AND BARBUDA
ST. KITTS AND NEVIS

UNITED STATES

BERMUDA
U.K.

DOMINICAN
REPUBLIC
BAHAMAS   PUERTO
                 RICO
MEXICO   U.S.

CUBA

CAYMAN IS.
U.K.   HAITI
BELIZE   JAMAICA
HONDURAS
GUATEMALA   NICARAGUA   TRINIDAD
EL SALVADOR   AND
               CURAÇAO   TOBAGO
COSTA RICA   Neth.
          PANAMA

Water Availability

(millimeters per person
per year)

- ■ More than 750
- ■ 251–750
- ■ 26–250
- □ Less than 26
- □ No data

BRITISH COLUMBIA
ALBERTA
SASKATCHEWAN
MANITOBA

Red Deer
Prince Albert
Lake Manitoba

Vancouver
Vancouver I.
Victoria
Cape Flattery
Kamloops
Calgary
Saskatoon
Regina
Moose Jaw
Weyburn
Brandon
Winnipeg

Kelowna
Medicine Hat
Lethbridge

Olympia
Seattle
Tacoma
WASH.
Spokane
Walla Walla
Havre
Great Falls
Missouri
Minot

Portland
Salem
Eugene
Coos Bay
Columbia
Redmond
Missoula
Bitterroot Range
MONTANA
Helena
Butte
Bozeman
Billings
Miles City

NORTH DAKOTA
Bismarck
Fargo

Medford
Klamath Falls
OREGON
IDAHO
Idaho Falls
Boise
Snake River
Pocatello
Absaroka Range
Bighorn Mts.
Worland
Aberdeen
SOUTH DAKOTA
Pierre

Eureka
Alturas
WYOMING
Black Hills
Rapid City
Sioux Falls

Sacramento
Reno
NEVADA
Salt Lake City
GREAT
Great Salt Lake
Provo
Casper
Laramie
Cheyenne
Sioux City
NEBRASKA
Omaha
Lincoln

San Francisco
Oakland
San Jose
Fresno
Carson City
SIERRA NEVADA
Ely
UTAH
BASIN
Boulder
Denver
Colorado Springs
Platte
KANSAS
Wichita

CALIFORNIA
Bakersfield
Santa Barbara
Channel Is.
LOS ANGELES
Death Valley
(-282 ft)
Mojave Desert
Las Vegas
Colorado
COLORADO
San Juan Mts.
Arkansas

Riverside
Long Beach
Grand Canyon
PLATEAU
Flagstaff
ARIZONA
PHOENIX
Santa Fe
Albuquerque
NEW MEXICO
Oklahoma City
OKLAHOM
Wichita Falls

San Diego
Tijuana
Ensenada
Yuma
Mexicali
Tucson
NEW MEXICO
Llano Estacado
Lubbock
Red
Amarillo

Nogales
Douglas
Carlsbad
El Paso
Hobbs
Dallas
Fort Worth
Waco

BAJA CALIFORNIA
Ciudad Juárez
Rio Bravo del Norte
Odessa
TEXAS
Edwards Plateau
Austin

GULF OF CALIFORNIA
Hermosillo
Chihuahua
San Antonio
Corpus Christi

PACIFIC OCEAN
Guaymas
MEXICO
Delicias
Nuevo Laredo
Laredo

La Purísima
Los Mochis
Monclova
Brownsville
Matamoros

TROPIC OF CANCER
Monterrey
Saltillo
Torreón
La Paz
Culiacán
Durango
Ciudad Victoria
Mazatlán

ROCKY MOUNTAINS
UNITED STATES
CASCADE RANGE
GULF OF CALIFORNIA

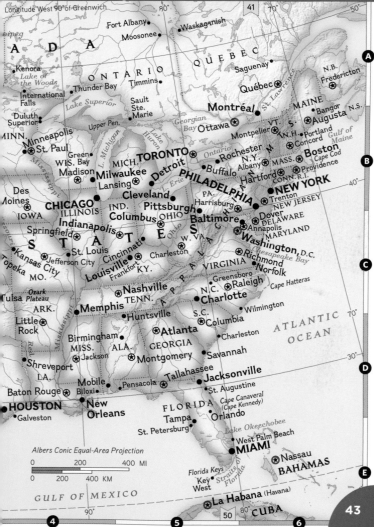

Longitude West 90° of Greenwich

CANADA

Fort Albany
Waskaganish
Moosonee

Kenora
Lake of the Woods
International Falls
Duluth
Superior
MINN.
Minneapolis
St. Paul
Green Bay
WIS.
Madison
Milwaukee

ONTARIO
Thunder Bay
Timmins
Sault Ste. Marie
Upper Pen.
Lake Superior
L. Michigan
MICH.
Lansing

QUEBEC
Saguenay
Québec
Montréal
Ottawa
St. Lawrence
Georgian Bay
Lake Huron
Lake Ontario
Rochester
N.Y.
Buffalo
Albany
MAINE
Bangor
Augusta
N.B.
Fredericton
N.S.
VT.
N.H.
Montpelier
Concord
Portland
Gulf of Maine
MASS.
Boston
Cape Cod
CONN. R.I.
Providence
Hartford

Des Moines
IOWA
CHICAGO
ILLINOIS
Springfield
IND.
Indianapolis
Columbus
OHIO
Detroit
TORONTO
Cleveland
Pittsburgh
PHILADELPHIA
PA.
Harrisburg
Lake Erie
Trenton
NEW YORK
NEW JERSEY

UNITED STATES

Kansas City
Topeka
MO.
Jefferson City
St. Louis
Cincinnati
Louisville
Frankfort
KY.
Charleston
W. VA.
Baltimore
Dover
DELAWARE
Annapolis
MARYLAND
Washington, D.C.
Richmond
VIRGINIA
Chesapeake Bay
Norfolk

Mississippi
Ozark Plateau
Tulsa
ARK.
Little Rock
Memphis
Nashville
TENN.
Huntsville
APPALACHIAN
N.C.
Greensboro
Raleigh
Charlotte
S.C.
Columbia
Wilmington
Cape Hatteras

Red
Shreveport
LA.
MISS.
Jackson
Birmingham
ALA.
Montgomery
Atlanta
GEORGIA
Savannah
Charleston
ATLANTIC OCEAN

Baton Rouge
HOUSTON
Galveston
Mobile
Biloxi
Pensacola
New Orleans
Tallahassee
Jacksonville
St. Augustine
FLORIDA
Cape Canaveral (Cape Kennedy)
Tampa
St. Petersburg
Orlando
Lake Okeechobee
West Palm Beach
MIAMI
Nassau
BAHAMAS

Albers Conic Equal-Area Projection
0   200   400 MI
0   200   400 KM

Florida Keys
Key West
Straits of Florida

GULF OF MEXICO

La Habana (Havana)
CUBA

A
B
C
D
E

4   5   6

43

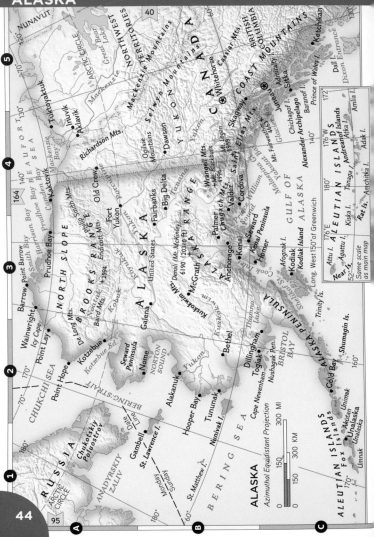

ALASKA
Azimuthal Equidistant Projection

0   150   300 MI
0   150   300 KM

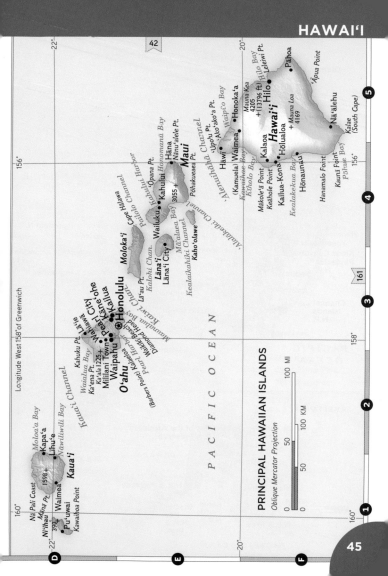

PRINCIPAL HAWAIIAN ISLANDS

Oblique Mercator Projection

Longitude West 158° of Greenwich

PACIFIC OCEAN

Kaua'i
Moloa'a Bay
Kapa'a
Līhu'e
Nāwiliwili Bay
Nā Pali Coast
Mana Pt. 1598+
Waimea
Pu'uwai
Kawaihoa Point
Ni'ihau

Kaua'i Channel

O'ahu
Kahuku Pt.
Waialua Bay
Ka'ena Pt.
Ka'a'la 1225+
Lā'ie
Wahiawā
Waipahu **Wai'anae** **Kāne'ohe**
Mililani Town **Kailua**
Pearl City
Waikīkī Beach ⊗ **Honolulu**
(Barbers Point) Kalaeloa Pt.
Diamond Head
Māmala Bay
Ka'ena Pt. Kaua'i Chan.

Moloka'i Chan.

Kaua'i Chan.

Moloka'i
Cape Hālawa
La'au Pt.

Kaiwi Chan.

Lāna'i
Lāna'i City
Keālaikahiki Channel

Kaho'olawe

Pailolo Channel
Kahului Harbor
'Opana Pt.
Wailuku **Kahului** **Hāna**
3055+ Nānu'alele Pt.
Mā'alaea Bay Pōhakueaea Pt.
Maui
Honomanū Bay

Keālaikahiki Channel

'Alalākeiki Channel

'Alenuihāhā Channel

Waipi'o Bay
'Upolu Pt.
Ako'ako'a Pt.
Hāwī **Waimea** **Honoka'a**
(Kamuela) Mauna Kea
Kawaihae Bay 4205 +13796 ft
Kholoho Bay
Māikole's Point
Keahole Point **Kalaoa** **Hōlualoa**
Kailua-Kona
Kealakekua Bay + Mauna Loa 4169
Hōnaunau
Hanamalo Point
Kauna Point
Kalae (South Cape)
Na'alehu
Pōhue Bay
Kalae

Hilo Bay Leleiwi Pt.
Pāhoa
'Āpua Point
Hilo
Hawai'i

42

161

45

85°   80°

Tampa
St. Petersburg
Sarasota

Orlando
Cape Canaveral
(Cape Kennedy)
Kissimmee

A

UNITED
STATES

Fort Pierce

West Palm Beach

Little
Abaco I. Cooper's Town

B

Charlotte Harbor

Fort Myers

Grand
Bahama I.   Freeport
Northwest
Providence
Channel

41

Abaco
Island

Southwest Point

Lake
Okeechobee   Fort
Lauderdale
MIAMI   Miami Beach

Cape Romano   Homestead

Bimini
Islands
Joulter
Cays

Berry
Islands

Eleuthera
Island

GULF   *Ponce de Leon Bay*
Cape Sable

New Providence

Nassau

Rock Sour

-25°

OF

Dry
Tortugas

Key
West

MEXICO

Williams I.
15

Andros
Island

Big Wood Cay

Flamin

Cat

F l o r i d a   K e y s

Kemps Bay

Cistern Pt.

Exuma
Sound

Tongue of
the Ocean

Cape Santa Maria
Great
Exuma   Little
Exuma

Deadmans Ca

B

S T R A I T S   O F   F L O R I D A

Nicholas Channel

La Habana
(Havana)

Matanzas

Archipiélago de Sabana

Old Bahama Channel

Pinar del Río
Bahía
Guadiana
Cabo
Francés
Pta. Frances

Los Palacios
Artemisa

+692

Güines

Península
de Zapata

CUBA

Santa Clara
Cienfuegos

Archipiélago de Camagüey

Morón

C

Golfo de
Batabanó

Nueva Gerona
+310   *Isla de la Juventud*
(Isle of Youth)
Punta del Guanal
Archipiélago de los Canarreos

Sancti
Spíritus

Ciego de
Ávila

Camagüey

Golfo de Ana María
Pta. Macurijes

San
Pedro

Las Tunas

Holguín

+123

Cabo de San Antonio

Jardines de la Reina

Guayabal
Cauto

Bayamo +

Yucatan Channel

-20°

Golfo de Guacanayabo

Manzanillo

Sierra Maestra

Santiago de Cuba

Pico +2005

85°

Little
Cayman
Cayman
Brac
CAYMAN ISLANDS
U.K.

Cabo
Cruz *Turquino*

Guantánamo
NAVA
BAS

GUANTANAM
BA

D

George Town
Grand
Cayman

49

E

Islas Santanilla
(Swan Islands)
Honduras

G
R
E
A
T
E
R

C
A
R
I
B

Montego
Bay

Montego Bay

Saint Ann's Bay   Port Antonio

North Negril Point
South Negril Point

Blue Mt. Pk.
JAMAICA

Northeast
Point   Morant Poin

Savanna-la-Mar

Great Pedro Bluff

May Pen
Spanish Town

Kingston

Portland Point

Pedro Cays
Jamaica

B

B

E

A

N

HONDURAS
Laguna de Caratasca
Cabo Falso

49

1   2   3

ATLANTIC OCEAN

Azimuthal Equidistant Projection

0   100   200 MI

0   100   200 KM

A   A   A   A   A   S

Point

San Salvador
(Watling)

Rum Cay

TROPIC OF CANCER

Long I.   Samana Cay

Crooked Island Passage

Crooked I.
61

Acklins I.

Mayaguana Passage   40

Mayaguana I.

Caicos Passage

TURKS AND
CAICOS ISLANDS
U.K.

Little Inagua I.   Caicos
Islands   Cockburn Town

Great Inagua   Ambergris   Turks Is.
Island   Cays

le Nipe   41
Mathew Town

Turks Island Passage

Mouchoir Passage

Punta Guarico

†175   Silver Bank Passage

Punta   Île de
de Maisí   la Tortue   Monte Cristi   Puerto Plata

Windward Passage   Cap-Haïtien   Cabo Francés Viejo

902   Port-   Gonaïves   Santiago   San Francisco de Macorís
Cap-   de-Paix   Cordillera   Bahía de   PUERTO
Foux   Central   La Vega   Samaná   RICO
Golfe de   †1788   Samaná   United States
la Gonâve   Pico Duarte   Bahía de Samaná
3175   Cabo   San Juan
Pointe Ouest   HISPANIOLA   Engaño
Île de la Gonâve   Arecibo   Culebra
Massif de   Neiba   Mona   †1338   Vieques
la Hotte   Jacmel   SANTO DOMINGO   Passage   Ponce
Cap   Pòtoprens   San Pedro de Macorís   Isla Mona   Cabo Rojo
Carcasse   Les   (Port-au-Prince)   Baní   Isla Saona
Pointe-à-   Cayes   †1605   Bahía de
Gravois   Barahona   Ocoa   52
Cabo   Bahía de Neiba
Falso   Cabo Beata
DOMINICAN
REPUBLIC

HAITI

T   I   L   L   E   S

A   N   S   E   A

A

68

Lake
Maracaibo

SURINAME

Bogotá

Amazon

Purus

B R A Z

Lima

Lake
Titicaca

Paraná

Cerro Aconcagua
6,959 m
(22,831 ft)

Buenos
Aires

Laguna del Carbón
-105 m
(-344 ft)

SOUTH AMERICA

GEOGRAPHIC EXTREMES

CONTINENTAL POLITICAL FACTS

TOTAL NUMBER OF COUNTRIES: 12

LARGEST COUNTRY BY AREA: Brazil
8,515,770 sq km (3,287,956 sq mi)

SMALLEST COUNTRY BY AREA:
Suriname 163,820 sq km
(63,251 sq mi)

MOST POPULOUS COUNTRY: Brazil
204,260,000

LEAST POPULOUS COUNTRY: Suriname
580,000

LARGEST URBAN AREAS BY POPULATION:
São Paulo, Brazil 21,066,000
Buenos Aires, Argentina 15,180,00
Rio de Janeiro, Brazil 12,902,000
Lima, Peru 9,897,000
Bogotá, Colombia 9,765,000

CONTINENTAL PHYSICAL FACTS

AREA: 17,819,000 sq km (6,880,000 sq mi)

HIGHEST POINT: Cerro Aconcagua,
Argentina 6,959 m (22,831 ft)

LOWEST POINT: Laguna del Carbón,
Argentina -105 m (-344 ft)

LONGEST RIVERS:
Amazon 6,680 km (4,150 mi)
Paraná-Río de la Plata 4,000 km
(2,490 mi)
Purus 3,380 km (2,100 mi)

LARGEST NATURAL LAKES (SURFACE AREA):
Lake Maracaibo *(recognized by some as a
lake)* 13,200 sq km (5,100 sq mi)
Lake Titicaca 8,300 sq km
(3,200 sq mi)

I L

Rio de Janeiro

São Paulo

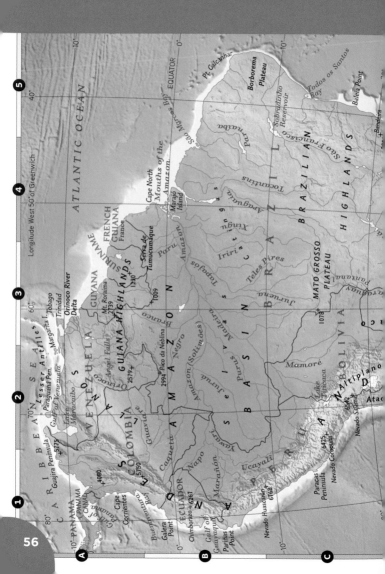

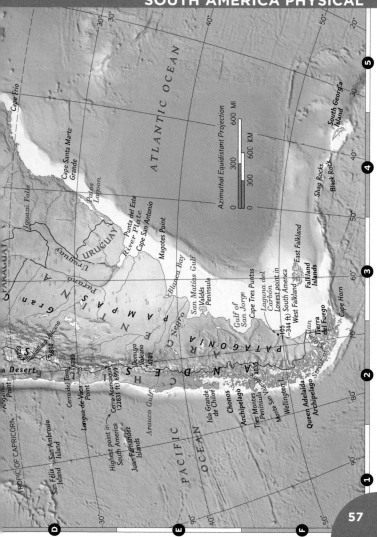

ATLANTIC OCEAN

Cape Frio

Cape Santa Marta Grande

Patos Lagoon

Iguazú Falls

URUGUAY

Paraná

Punta del Este
River Plate
Cape San Antonio

Magotes Point

Blanca Bay

San Matías Gulf
Valdés Peninsula

Gran
PARAGUAY
Paraguay

PAMPAS

Negro

Gulf of
San Jorge
Cape Tres Puntas

Laguna del
Carbón
-105
(-344 ft)
Lowest point in
South America

South Georgia
Island

Shag Rocks
Black Rock

East Falkland

Falkland
Islands

West Falkland

Cape Horn

Azimuthal Equidistant Projection

600 MI
300

60C KM
300
0

Anganco
Point

S

5723
6880
6380

Domuyo
Volcano
4709

Cerro del Toro

Langua de Vaca
Point

Cerro Aconcagua
(22831 ft) 6959

Arauco Gulf

a Desert

San Félix
Island

San Ambrosio
Island

Highest point in
South America

Juan Fernández
Islands

TROPIC OF CAPRICORN

ANDES

PATAGONIA

Isla Grande
de Chiloé

Chonos
Archipelago

Tres Montes
Peninsula

Monte San Valentín 4035

Wellington I.

Queen Adelaida
Archipelago

Tierra
del Fuego

PACIFIC
OCEAN

Climate Zones

(based on modified Köppen system)

Humid equatorial climate (A)
- No dry season (Af)
- Short dry season (Am)
- Dry winter (Aw)

Dry climate (B)
- Semiarid (BS) } h = hot
- Arid (BW)   } k = cold

Humid temperate climate (C)
- No dry season (Cf) }
- Dry winter (Cw)   } a = hot summer
- Dry summer (Cs)   } b = cool summer

Cold climate (E)
- Tundra and ice

Highland climate (H)
- Unclassified highlands

VENEZUELA
GUYANA
SURINAME
FRENCH GUIANA Fr.
COLOMBIA
ECUADOR
GALÁPAGOS
ISLANDS
Ecuador
PERU
BRAZIL
BOLIVIA
PARAGUAY
CHILE
URUGUAY
ARGENTINA
FALKLAND
ISLANDS
U.K.

Water Availability

(millimeters per person
per year)

- More than 750
- 251–750
- 26–250
- Less than 26

EUROPE

GEOGRAPHIC EXTREMES

CONTINENTAL POLITICAL FACTS

TOTAL NUMBER OF COUNTRIES: 46

LARGEST COUNTRY BY AREA: Russia
17,098,242 sq km (6,601,665 sq mi)

SMALLEST COUNTRY BY AREA:
Vatican City   0.44 sq km (0.17 sq mi)

MOST POPULOUS COUNTRY: Russia
142,355,000

LEAST POPULOUS COUNTRY: Vatican City
1,000

LARGEST URBAN AREAS BY POPULATION:
Istanbul, Turkey   14,164,000
Moscow, Russia   12,166,000
Paris, France   10,843,000
London, United Kingdom   10,313,000
Madrid, Spain   6,199,000

CONTINENTAL PHYSICAL FACTS

AREA: 9,947,000 sq km (3,841,000 sq mi)

HIGHEST POINT: El'brus, Russia   5,642 m
(18,510 ft)

LOWEST POINT: Caspian Sea   -28 m
(-92 ft)

LONGEST RIVERS:
Volga   3,690 km (2,290 mi)
Danube   2,860 km (1,780 mi)
Dnieper   2,290 km (1,420 mi)

LARGEST NATURAL LAKES:
Caspian Sea   371,000 sq km
(143,200 sq mi)
Lake Ladoga   17,700 sq km
(6,800 sq mi)
Lake Onega   9,800 sq km
(3,800 sq mi)

London

Paris

Madrid

VATICAN
CITY

Lake Onega

Lake Ladoga

RUSSIA

Volga

Moscow

Kaliningrad (Russia)

Dnieper

Caspian Sea
(-92 ft) -28 m
See p. 79

Danube

İstanbul

El'brus
5,642 m
(18,510 ft)

Map Labels

Projection / Scale
Azimuthal Equidistant Projection

0 — 200 — 400 MI
0 — 200 — 400 KM

A commonly accepted division between Asia and Europe—here marked by a green line—is formed by the Ural Mountains, Ural River, Caspian Sea, Caucasus Mountains, and the Black Sea with its outlets, the Bosporus and Dardanelles.

Water bodies
NORWEGIAN SEA
ATLANTIC OCEAN
NORTH SEA
CELTIC SEA
BAY OF BISCAY
English Channel
Irish Sea
BALTIC
GULF OF
TYRRHENIAN SEA
IONIAN SEA
ADRIATIC SEA
Faxaflói
Húnaflói
Þistilfjörður
Vestfjorden
Moray Firth
Skagerrak
Golfe du Lion
MEDITERRANEAN
Strait of Gibraltar

Iceland / North Atlantic
Hólmavík
Akureyri
ICELAND
Langanes
Vopnafjörður
Reykjavík
Höfn
Tórshavn
Faroe Islands
Denmark
Shetland Is.
Lerwick
Orkney Is.
Isle of Lewis
Wick
Inverness

Scandinavia / Norway / Sweden
Tromsø
Narvik
Svolvær
Bodø
Mo i Rana
Namsos
Trondheim
Örnsköldsvik
Lillehammer
Bergen
Gävle
Skien
Kristiansand
Oslo
Stockholm
Göteborg
Gotland
NORWAY
SWEDEN
SCANDINAVIA
Meridian of Greenwich (London)

United Kingdom / Ireland
Belfast
Glasgow
Edinburgh
UNITED KINGDOM
Dublin
(Baile Átha Cliath)
(ÉIRE) IRELAND
(Sionainn) Shannon
Cork (Corcaigh)
Cardiff
Birmingham
Manchester
LONDON
Land's End
U.K. Channel Is.
Brest
Pointe de Saint-Mathieu

Denmark / Germany / Central Europe
DENMARK
København (Copenhagen)
Hamburg
Gdańsk
POLAND
Poznań
Łódź
Berlin
GERMANY
Amsterdam
NETH.
Bruxelles (Brussels)
BELG.
LUX.
Praha (Prague)
CZECHIA (CZECH REP.)
Bratislava
München (Munich)
Wien (Vienna)
Elbe

France / Alps / Italy
Nantes
Rennes
PARIS
Strasbourg
Bordeaux
Lyon
Toulouse
FRANCE
SWITZ.
Bern
LIECH.
AUSTRIA
SLOV.
Massif Central
Pyrénées
ANDORRA
Marseille
MONACO
SAN MARINO
VATICAN CITY
Milano (Milan)
Po
Ljubljana
Zagreb
CROAT.
BOSN. & HERZG.
Sarajevo
Podgorica
MONT.
Tiranë (Tirana)
ALBANIA
SLOV. HUNG.
Roma (Rome)
Napoli (Naples)
ITALY
Corsica
Sardinia
Cagliari
Palermo
Sicily
Catania
Valletta
MALTA
Loire
Seine
Rhône
A Coruña
Vigo
(Oporto) Porto
Coimbra
PORTUGAL
Lisboa (Lisbon)
Setúbal
Bilbao
Valladolid
Pamplona
Cabo de São Vicente

Iberia / North Africa
MADRID
BARCELONA
SPAIN
Sevilla
Córdoba
Cádiz
Granada
Valencia
Cartagena
Cap de Tortosa
Balearic Islands
(Tangier) Tanger
GIBRALTAR U.K.
Ceuta Sp.
Casablanca
Rabat
Fès (Fez)
MOROCCO
ATLAS MOUNTAINS
Melilla Sp.
Oran
Alger (Algiers)
ALGERIA
Tunis
TUNISIA
Tagus
Longitude West of Greenwich
Longitude East of Greenwich

ICELAND

Faroe Islands
Denmark

NORWAY

SWEDEN

FINLAND

EST

IRELAND

DENMARK

UNITED
KINGDOM

NETH.

RUSS.

LITH

London

POLAND

Channel Is.
U.K.

BELG.

GERMANY

Paris

LUX.

LIECH

CZECHIA
(CZECH REP.)

SLOVAKIA

FRANCE

SWITZ.

AUSTRIA

HUNGARY

SLOV.

CROATIA

MONACO

BOSN. &
HERZG.

SERBIA

PORTUGAL

ANDORRA

SAN MARINO

Madrid

ITALY

MONTENEGRO

KOS.

MACED.

SPAIN

Barcelona

VATICAN
CITY

ALBANIA

GREEC

Gibraltar
U.K.

84

MALTA

Land Cover

- ■ Evergreen needleleaf forest
- ■ Evergreen broadleaf forest
- ■ Deciduous needleleaf forest
- ■ Deciduous broadleaf forest
- ■ Mixed forest
- ■ Woody savanna
- ■ Savanna
- ■ Closed shrubland

- ☐ Open shrubland
- ■ Grassland
- ☐ Cropland
- ■ Barren or sparsely vegetated
- ■ Urban or built-up
- ☐ Snow and ice
- ■ Cropland/natural vegetation mosaic
- ■ Wetland

- O Urban area with more than 5 million inhabitants

85

Climate Zones

(based on modified Köppen system)

Dry climate (B)

☐ Semiarid (BS)
☐ Arid (BW) } k = cold

Humid temperate climate (C)

■ No dry season (Cf)
☐ Dry summer (Cs)

Humid cold climate (D)

☐ No dry season (Df)

Cold climate (E)

■ Tundra and ice

Highland climate (H)

☐ Unclassified highlands

a = hot summer
b = cool summer
c = short, cool summer

Water Availability

(millimeters per person per year)

- ■ More than 750
- ■ 251–750
- ☐ 26–250
- ☐ Less than 26
- ☐ No data

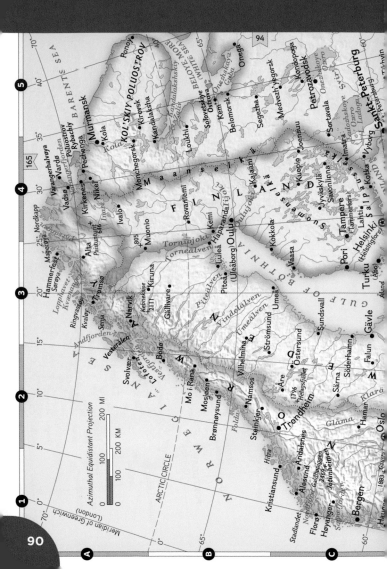

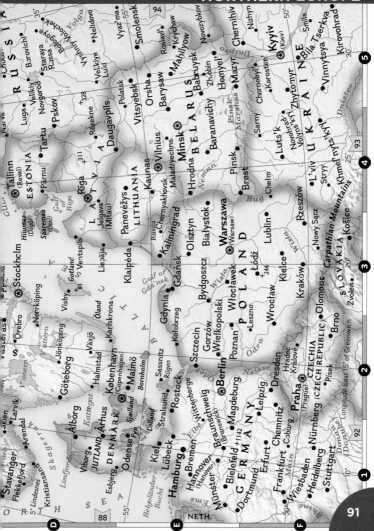

Map labels (reading across the map):

RUSSIA — Volga, Vyšnij Voločëk, Nelidovo, Vyaz'ma, Smolensk, Bologoye, Borovichi, Staraya Russa, Velikiy Novgorod, Valdiy Luki, Roslavl', Mahilyow, Krychaw, Novozybkov, Chernihiv, Nizhyn, Smila, Kyiv (Kiev), Luga, Pskov, Polatsk, Vitsyebsk, Orsha, Barysaw, Babruysk, Homyel', Mazyr, Korosten, Zhytomyr, Bila Tserkva, Kirovohrad, Vinnytsya, Nava, Tallinn (Reval), Tartu, Razekne, Daugavpils, Minsk, Baranavichy, Pinsk, Sarny, Luts'k, Novovad, Zhytomyr

ESTONIA — Hiiumaa (Dagö), Pärnu, Saaremaa (Ösel), Gulf of Riga

LATVIA — Rīga, Jelgava (Mitau), Panevėžys, +311

LITHUANIA — Vilnius, Kaunas, Hrodna, Maladzyechna, Malvdzyechna, Neman

BELARUS — Minsk, Vitsyebsk, Barysaw, Babruysk, Homyel', Brest, Pinsk Marshes, Pinsk

UKRAINE — L'viv, Stryy, Khmel'nyts'kyy, Chernivtsi, Dnister, Carpathian Mountains, Nowy Sącz, Košice, Zvolen

SLOVAKIA — Košice, Zvolen

Klaipėda, Chernyakhovsk, Kaliningrad, Białystok, Olsztyn, Warszawa (Warsaw), Lublin, Rzeszów, Chelm, Bug, Russia

POLAND — Gdańsk, Gdynia, Bydgoszcz, Włocławek, Łódź, Kielce, Kraków, Wrocław, Leszno, Poznań, Gorzów Wielkopolski, Szczecin, Kołobrzeg, Gulf of Gdańsk, Wisła, +246

Stockholm, Norrköping, Visby, Gotland, Öland, Karlskrona, Växjö, Jönköping, Örebro, Göteborg, Vänern, Vättern, Halmstad, Bornholm, Rügen, Sassnitz

DENMARK — København (Copenhagen), Malmö, Odense, Fyn, Sjælland, Ålborg, Århus, Viborg, JUTLAND, Esbjerg, Kiel, Kattegat, Limfjorden, Skagerrak

GERMANY — Berlin, Magdeburg, Braunschweig, Hannover (Hanover), Bremen, Hamburg, Lübeck, Rostock, Stralsund, Wittenberge, Leipzig, Dresden, Chemnitz, Erfurt, Coburg, Nürnberg, Stuttgart, Heidelberg, Wiesbaden, Frankfurt, Main, Bielefeld +1142, Münster, Dortmund, Ems, Brunswick, Schweig

CZECHIA (CZECH REPUBLIC) — Praha (Prague), Brno, Olomouc, Hradec Králové, Pisek, Rhine, Danube

Stavanger, Skien, Larvik, Arendal, Kristiansand, Flekkefjord, Lindesnes, NORTH SEA, Helgoländer Bucht, Neisse, Odra, Elbe

Longitude East 15° of Greenwich

NETH.

Grid references: 94, 93, 5, 4, 3, 2, 1, D, E, F, 88

Karlsruhe
Strasbourg
FRANCE
Freiburg
(Prague) Praha
GERMANY
Nürnberg
Stuttgart
Augsburg
Písek
10°
15°
CZECHIA
(CZECH REPUBLIC)
Brno
Olomouc
Kraków
20°
Rzeszów
POLAND
2655
Passau
Danube
Wien
(Vienna)
SLOVAKIA
Zvolen
Košice
Eger
München
(Munich)
Zürich
Bern
SWITZERLAND
Lausanne
+4478
Linz
St. Pölten
Bratislava
(Pressburg)
Innsbruck
LIECH.
3772
Salzburg
AUSTRIA
Leoben
Graz
Wiener
Neustadt
Mátra
1015
Budapest
HUNGARY
Debrec
Chur
Brenner Pass
Belluno
Venice
Udine
Klagenfurt
Maribor
Zalaegerszeg
Balaton
Szolno
Szeged
Orade
Milano
(Milan)
Verona
Venezia
SLOVENIA
Ljubljana
Drava
Mures
Timișoara
Po
Piacenza
Po
Gulf of
Venice
Rijeka
Zagreb
Osijek
Novi
Sad
Beograd
(Belgrade)
Genova
(Genoa)
Bologna
Sava
Dunav
La Spezia
MONACO
Livorno
Firenze
(Florence)
SAN
MARINO
Pesaro
Zadar
Banja Luka
Tuzla
BOSNIA AND
HERZEGOVINA
Sarajevo
Valjevo
SERBIA
Cap Corse
Bastia
CORSICA
France
Monte Cinto
2710
Ajaccio
Elba
Viterbo
Perugia
ITALY
2912+
Roma
(Rome)
VATICAN CITY
Split
Mostar
2522
Podgorica
Nikšić
MONTENEGRO
Dubrovnik
Niš
Prishtinë
(Pristina)
KOSOVO
Skopj
Porto-Vecchio
Olbia
Sassari
Nuoro
SARDINIA
Italy
Oristano
1834
Cagliari
Capo Comino
Napoli
(Naples)
Vesuvio
1281
Salerno
Foggia
Bari
Brindisi
Taranto
Lecce
Skadarsko Jezero
Shkodër
Tiranë
(Tirana)
ALBANIA
Berat
Korçë
Prizren
MACEDO
Prilep
Bito
Véria
Olimbo
(Olympus
Capo Carbonara
Punta Licosa
Golfo di
Taranto
Sapri
Strait of
Otranto
Kérkira
(Corfu)
Ioánina
Larisa
Lam
Capo Spartivento
Golfo di
Policastro
Punta Alice
Cosenza
Catanzaro
Capo Vaticano
AGrínio
Kefaloniá
Pátra
(Patrae)
Pirgo
Bizerte
Tunis
TUNISIA
Nabeul
Qairouan
Gulf of
Hamamet
Sousse
Mahdia
Sfax
Gulf of Gabes
Palermo
Agrigento
SICILY
Modica
MALTA
Valletta
Messina
Etna
3350
Reggio di Calabria
Stretto di Messina
Catania
Siracusa
Capo
Passero
IONIAN SEA
Zákinthos
PELOPONNES
Kalamáta
240
Akroti
Téna
Azimuthal Equidistant Projection
0   100   200 MI
0   100   200 KM
Longitude East 15° of Greenwich
20°

139

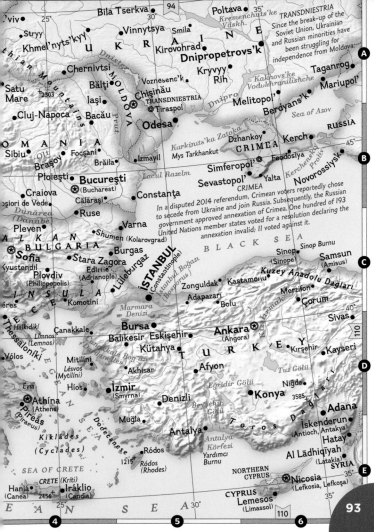

'viv 25°

Bila Tserkva 94 Poltava 35° TRANSDNIESTRIA
Stryy Vinnytsya Smila Kremenchats'ke Since the break-up of the
Khmel'nyts'kyy Kirovohrad Vdskh. Soviet Union, Ukrainian
Chernivtsi UKRAINE Dnipropetrovs'k and Russian minorities have
Bălţi Kryvyy been struggling for
Satu Iaşi Voznesens'k Rih Kakhovs'ke Taganrog independence from Moldova.
Mare +2303 Chişinău Vodulvranilishche Mariupol'
Cluj-Napoca Bacău TRANSDNIESTRIA Dnipro Melitopol'
Tiraspol Berdyans'k Sea of Azov
ROMANIA Odesa RUSSIA
Sibiu Focşani Karkinits'ka Zatoka Dzhankoy Kerch
Braşov Brăila Izmayil Mys Tarkhankut CRIMEA 45°
Ploieşti Simferopol' Feodosiya Kerchens'ka
Craiova Bucureşti Lacul Razelm Sevastopol' Yalta Protis Novorossiysk
şori de Vede (Bucharest) CRIMEA
Călăraşi Constanţa In a disputed 2014 referendum, Crimean voters reportedly chose
Dunărea Ruse to secede from Ukraine and join Russia. Subsequently, the Russian
Pleven (Danube) Varna government approved annexation of Crimea. One hundred of 193
BALKAN Shumen (Kolarovgrad) United Nations member states voted for a resolution declaring the
BULGARIA Burgas annexation invalid; 11 voted against it.
Sofia Stara Zagora BLACK SEA
yustendil Plovdiv Edirne Lüleburgaz Sinop Sinop Burnu Samsun
(Philippopolis) (Adrianople) İSTANBUL Kuzey Anadolu Dağları (Amisus)
INSULA Komotiní Constantinople Zonguldak Kastamonu Merzifon Çorum
éres Halkidikí Marmara İstanbul Boğazı Adapazarı Çorum
Thessaloníki Limnos Denizi (Bosporus) Bolu Sivas
Vólos (Lemnos) Çanakkale Bursa Eskişehir Ankara 40°
Evía Balıkesir Kütahya (Angora) Kırşehir Kayseri
Mitilíni Akhisar Afyon TURKEY
Athína Lésvos İzmir Niğde Adana
(Athens) (Mytilíni) (Smyrna) Eğridir Gölü Tuz Gölü İskenderun
Pireás Híos Denizli Konya 3585 (Antioch, Antakya)
(Piraeus) Beyşehir Hatay
Kikládes Mugla Gölü Toros Dağları Al Lādhiqīyah
(Cyclades) Ródos Antalya (Latakia)
SEA OF CRETE (Rhodes) Antalya NORTHERN SYRIA
1215 Ródos Körfezi CYPRUS Nicosia 35°
Haniá CRETE (Kríti) Yardımcı (Lefkosia, Lefkoşa)
(Canea) 2456 İráklio Burnu CYPRUS
EAN (Candia) SEA 30° Lemesos 110
25° 4 5 (Limassol) 6 **93**

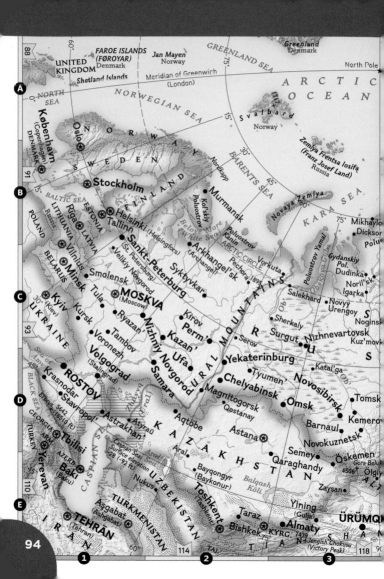

ARCTIC OCEAN

North Pole

GREENLAND SEA

Greenland
Denmark

FAROE ISLANDS
(FØROYAR)
Denmark

Jan Mayen
Norway

UNITED
KINGDOM

Shetland Islands

Meridian of Greenwich
(London)

A

NORTH
SEA

NORWEGIAN SEA

Svalbard
Norway

Zemlya Frentsa Iosifa
(Franz Josef Land)
Russia

København
(Copenhagen)
DENMARK

Oslo

N
O
R
W
A
Y

S
W
E
D
E
N

Nordkapp

BARENTS SEA

Novaya Zemlya

KARA SEA

B

Stockholm

BALTIC SEA

FINLAND

Murmansk

Kol'skiy
Poluostrov

Mikhaylo

Dickson

Polu

Helsinki (Helsingfors)

Tallinn

ESTONIA

LATVIA

Riga
Russia

Beloye More
(White Sea)

Arkhangel'sk
(Archangel)

Poluostrov
Pechora 1895

ARCTIC CIRCLE

Vorkuta

Poluostrov Yamal

Gydanskiy
Pol.

Dudinka

Noril'sk

Igarka

Ostrova Gyda

LITHUANIA

Vilnius

POLAND

BELARUS

Minsk

Sankt-Peterburg
(St. Petersburg)

Velikiy Novgorod

Syktyvkar

Salekhard

Novyy
Urengoy

Sherkaly

Smolensk

MOSKVA
(Moscow)

Kirov

Perm'

U
R
A
L

M
O
U
N
T
A
I
N
S

R

Serov

Surgut

Nizhnevartovsk

Kuz'mov

Noginsk

C

Kyiv
(Kiev)

UKRAINE

Tula

Ryazan'

Kursk

Tambov

Nizhniy Novgorod

Kazan'

Ufa

Yekaterinburg

Tyumen'

U

S

Katal'ga

Voronezh

Samara

Chelyabinsk

Novosibirsk

Tomsk

Volgograd
(Stalingrad)

Magnitogorsk

Omsk

Krasnodar

ROSTOV

Stavropol'

El'brus 5642
(18510 ft)

Astrakhan'

Aqtöbe

Qostanay

Barnaul

Kemero

GEORGIA

Tbilisi

Atyraū

Astana

Semey

Novokuznetsk

Öskemen

Gora Belu
4506

Ölgiy

D

BLACK SEA

Sea of
Azov

TURKEY

AZERB.

ARM.

Yerevan

Baki
(Baku)

CASPIAN SEA

Caspian Sea: lowest
surface elevation
-28 m (-92 ft)

Aral
Sea

Aral

K
A
Z
A
K
H
S
T
A
N

Qaraghandy

Balqash
Köli

Zaysan

E

IRAN

TEHRĀN
(Tehran)

Aşgabat
(Ashgabat)

TURKMENISTAN

Nukus

UZBEKISTAN

Bayqongyr
(Baykonur)

Toshkent
(Tashkent)

Taraz

Bishkek

KYRG.

Almaty

Yining
(Gulja)

ÜRÜMQ

SHAN

Jengish Chokusu
(Victory Peak)
7439

T
I
A
N

1   114   2   118   3

90

91

93

110

A commonly accepted division between Asia and Europe—here marked by a green line—is formed by the Ural Mountains, Ural River, Caspian Sea, Caucasus Mountains, and the Black Sea with its outlets, the Bosporus and Dardanelles.

Two-Point Equidistant Projection

300   600 MI
300   600 KM

ATLANTIC OCEAN

GREENLAND SEA
Greenland
Iceland
BRITISH ISLES
Great Britain
NORWEGIAN SEA
Svalbard
BARENTS SEA
Novaya Zemlya

SCANDINAVIA
Kola Pen.
White Sea
G. of Bothnia
Baltic Sea
NORTHERN EUROPEAN PLAIN

EUROPE
Volga
THE STEPPES
Crimea
Caspian Depression
Caspian Sea
Elbrus 5642
URAL MOUNTAINS

MOROCCO
ATLAS MOUNTAINS
ALGERIA
TUNISIA
MEDITERRANEAN SEA
IBERIAN PENINSULA
BALKAN PENINSULA
ANATOLIA (ASIA MINOR)
Cyprus
SYRIA
LIBYA
EGYPT
SAHARA
AFRICA
CHAD
SUDAN
RED SEA
ETHIOPIAN HIGHLANDS
ETHIOPIA
SOMALIA
Somali Peninsula
UGANDA
KENYA
Lake Victoria
Kilimanjaro 5895
TANZANIA
MOZAMBIQUE
COMOROS
Comoro Is.
Madagascar
Seychelles
SEYCHELLES

SYRIAN DESERT
MESOPOTAMIA
IRAQ
IRAN
Zagros Mts.
Persian Gulf
Gulf of Oman
SAUDI ARABIA
ARABIAN PENINSULA
Rub' al Khali
YEMEN
OMAN
Gulf of Aden
Socotra
Ra's al Hadd
BAHRAIN
QATAR
Gulf of Oman
ARABIAN SEA
Lakshadweep
MALDIVES
Maldive Islands
Chagos Archipelago
EQUATOR
INDIAN OCEAN

KAZAKH UPLANDS
KAZAKHSTAN
Turan Lowland
TURKMENISTAN
UZBEKISTAN
Aral Sea
Tian Shan
Taklimakan Desert
Kunlun
HINDU KUSH 7649
AFGHANISTAN
PAKISTAN
K2 (Godwin Austen) 8611
HIMALAYAS
Mount Everest 29035, World's highest point
Great Indian Desert
Ganges
INDIA
DECCAN PLATEAU
Western Ghats
Eastern Ghats
Cape Comorin
SRI LANKA
NEPAL
BHUTAN
BANGLADESH
MYANMAR

SIBERIA
Severnaya Zemlya (North Land)
+935
Ostrov Vil'kitskogo
Mys Dika (New Siberian Is.)
Novosibirskiye Ostrova
New Siberian Is. +374
LAPTEV SEA
EAST SIBERIAN SEA
CHUKCHI SEA
Bering Strait
St. Lawrence I. U.S.
Ostrov Vrangelya (Wrangel I.)
Proliv Longa
Date Line
Chukotskiy Poluostrov
Mys Navarin
BERING SEA
+1843
Mys Ozernyy
Komandorskiye Ostrova
Koryaki Mys Lopatka
Poluostrov Kamchatka (Kamchatka Pen.)
+1750
Petropavlovsk-Kamchatskiy
+1830
Klyuchi
SEA OF OKHOTSK
Kuril Islands
Severo Kuril'sk
Ostrov Sakhalin
Yuzhno-Sakhalinsk
Sovetskaya Gavan'
La Pérouse Strait
Hokkaido +2290
Sapporo
JAPAN
Honshu
Kyoto
SEA OF JAPAN (EAST SEA)
Vladivostok
SOUTH KOREA
SEOUL
NORTH KOREA
P'yongyang
YELLOW SEA
Bo Hai
TIANJIN
BEIJING
Tangshan
Baotou
Hami
CHINA
GOBI
MONGOLIA MOUNTAINS
Ulaanbaatar (Ulan Bator)
Erdenet
Bayanhongor
Mandalgovi
+3957
+3802
+2029
Da Hinggan Ling
Ulanhot
Yakeshi
Qiqihar
Harbin
Changchun
Jilin
Fushun
SHENYANG
Anshan
Dalnegorsk
Vyazemskiy
Khabarovsk
Komsomol'sk na Amure
Ushumun
Blagoveshchensk
Tynda
Chul'man
Novyy Uoyan
Ust' Ilimsk
Bodaybo
Vitim
Yerema
Bratsk
Krasnoyarsk
Angarsk
Irkutsk
Ulan Ude
Chita
Ozero Baykal (Lake Baykal)
Severo Yeniseyskiy
Mutoray
Tura
Kislokan
Yeniseyskiy
Mirnyy
Olekminsk
+2412
Aldan
Chagda
Chumikan
Okha
Okhotsk
Ust' Maya
Yakutsk
Orto Surt
Udachnyy
Chirinda
Put' Lenina
Boyarka +656
Zhilinda
Natara
Verkhoyanskiy Khrebet
Khatanga
Ust' Olenek
Tiksi
Aryy
Mys Taymyr
Nizhneyansk
Kazach'ye
Deputatskiy
Zyryanka
Belaya Gora
Srednekolymsk
Chersky
Bilibino
Fevek
Pevek
Omolon
Evensk
Manily +2562
Beringovskiy
Anadyr'
Mys Dezhneva
Mys Shmidta
Khrebet Cherskogo
Oymyakon
Gora Mus-Khaya +2959
+2374
Verkhoyansk
Magadan
Tukchi
Mys Yelizavety
Severo Kuril'sk
Lena
ARCTIC CIRCLE
Longitude East 105° of Greenwich
Longitude East 60° of Greenwich
Sunday / Monday

GEOGRAPHIC EXTREMES

CONTINENTAL POLITICAL FACTS

TOTAL NUMBER OF COUNTRIES: 46

LARGEST COUNTRY BY AREA: China
9,596,960 sq km (3,705,405 sq mi)

SMALLEST COUNTRY BY AREA:
Maldives   298 sq km (115 sq mi)

MOST POPULOUS COUNTRY: China
1,367,485,000

LEAST POPULOUS COUNTRY:
Maldives   393,000

LARGEST URBAN AREAS BY POPULATION:
Tokyo, Japan   38,001,000
Delhi, India   25,703,000
Shanghai, China   23,741,000
Mumbai (Bombay), India   21,043,000
Beijing, China   20,384,000

CONTINENTAL PHYSICAL FACTS

AREA: 44,570,000 sq km
(17,208,000 sq mi)

HIGHEST POINT: Mount Everest,
China-Nepal   8,850 m (29,035 ft)

LOWEST POINT: Dead Sea,
Israel-Jordan   -427 m (-1,401 ft)

LONGEST RIVERS:
Chang Jiang (Yangtze)   6,244 km
(3,880 mi)
Yenisey-Angara   5,810 km (3,610
Huang (Yellow)   5,778 km (3,59

LARGEST NATURAL LAKES:
Caspian Sea   371,000 sq km
(143,200 sq mi)
Lake Baikal   31,500 sq km
(12,200 sq mi)
Lake Balkhash   16,400 sq km
(6,300 sq mi)

Map labels:
Yenisey
Angara
kal (Baikal)
Caspian Sea
Lake Balkhash
Tokyo
Beijing
Shanghai
C H I N A
Huang (Yellow)
Chang Jiang (Yangtze)
Taiwan
Dead Sea
-427 m
(-1,401 ft)
Delhi
Mumbai
(Bombay)
Mount Everest
8,850 m
(29,035 ft)
MALDIVES

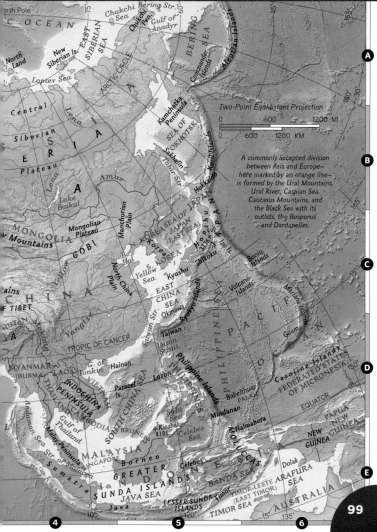

North Pole

Chukchi
Bering Str.

ARCTIC OCEAN

Chukchi
Sea

Chukchi
Pen.

Gulf of
Anadyr

North
Land

New
Siberian Is.

EAST
SIBERIAN
SEA

BERING SEA

Commander
Islands

Aleutian
Islands

A

Laptev Sea

ARCTIC CIRCLE

Central

Siberian

Plateau

S I B E R I A

Lena

Kamchatka
Peninsula

SEA OF
OKHOTSK

Two-Point Equidistant Projection

0   600   1200 MI

0   600   1200 KM

B

Lena

Amur

Sakhalin

Tatar Str.

Kuril Islands

Hokkaido

A commonly accepted division
between Asia and Europe—
here marked by an orange line—
is formed by the Ural Mountains,
Ural River, Caspian Sea,
Caucasus Mountains, and
the Black Sea with its
outlets, the Bosporus
and Dardanelles.

Lake
Baikal

MONGOLIA

Mongolian
Plateau

Manchurian
Plain

KOREA
N. KOREA

SEA OF JAPAN
(EAST SEA)

Honshu

JAPAN

Fuji
3776

C

Mountains

GOBI

Bo
Hai

Yellow
Sea

S. KOREA

Korea Str.

Kyushu

Shikoku

Bonin
Islands

CHINA

North China
Plain

EAST
CHINA
SEA

Okinawa

Ryukyu Islands

Volcano
Islands

Mariana
Islands

of TIBET

Yangtze

TROPIC OF CANCER

Taiwan

Luzon Strait

PHILIPPINE SEA

PACIFIC OCEAN

Guam

D

MYANMAR
BURMA

LAOS

VIETNAM

G. of
Tonkin

Hainan

Paracel
Is.

SOUTH CHINA SEA

Luzon

PHILIPPINES

Babelthuap
PALAU

Caroline Islands

FEDERATED STATES
OF MICRONESIA

INDOCHINA
PENINSULA

THAILAND
CAMBODIA

Gulf of
Thailand

Sulu
Sea

BRUNEI

Kinabalu
4101

Mindanao

Halmahera

EQUATOR

PAPUA
NEW
GUINEA

Andaman
Sea

Malay Peninsula

Str. of Malacca

SINGAPORE

MALAYSIA

Borneo

Celebes

Celebes
Sea

MOLUCCAS

NEW
GUINEA

E

GREATER
SUNDA ISLANDS

I N D O N E S I A

BANDA SEA

Dolak

ARAFURA
SEA

Sumatra

Java Sea

LESSER SUNDA
ISLANDS

Java

Celebes

Timor

TIMOR-LESTE
(EAST TIMOR)

TIMOR SEA

AUSTRALIA

105

4

120

5

135

6

Population Density

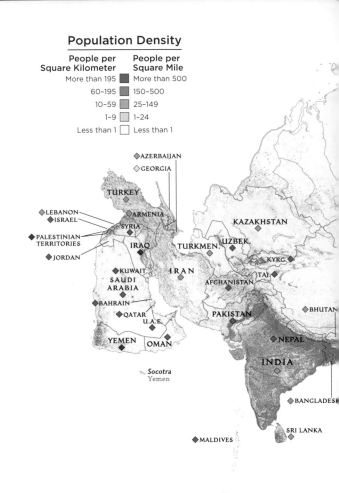

People per Square Kilometer	People per Square Mile
More than 195	More than 500
60–195	150–500
10–59	25–149
1–9	1–24
Less than 1	Less than 1

AZERBAIJAN

GEORGIA

TURKEY

LEBANON
ISRAEL

ARMENIA

SYRIA

KAZAKHSTAN

PALESTINIAN
TERRITORIES

JORDAN

IRAQ

TURKMEN.

UZBEK.

KYRG.

KUWAIT

I R A N

TAJ.

SAUDI
ARABIA

AFGHANISTAN

BAHRAIN

QATAR

U.A.E.

PAKISTAN

BHUTAN

YEMEN

OMAN

NEPAL

I N D I A

Socotra
Yemen

BANGLADESH

SRI LANKA

MALDIVES

Population Change

**Projected Population
Change, 2010–2050**

◆ More than 100%
◆ 50%–100%
◇ 0.01%–49%
◇ No change
◇ Population loss

RUSSIA ◇

JAPAN ◇

MONGOLIA ◇

N. KOREA ◇

S. KOREA ◇

CHINA

TAIWAN ◇

Hong Kong ◆
Macau ◆

MYANMAR
(BURMA)
LAOS ◆

PHILIPPINES ◆

THAILAND

VIETNAM ◇

CAMBODIA

BRUNEI ◆

New
Guinea

MALAYSIA ◆

Borneo

Sumatra

INDONESIA ◆

SINGAPORE ◇

Java

TIMOR-LESTE
(EAST TIMOR) ◆

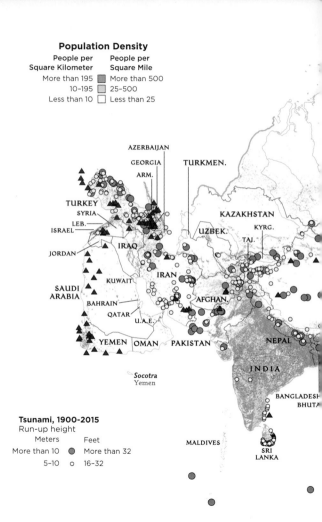

Population Density

People per Square Kilometer		People per Square Mile
More than 195		More than 500
10–195		25–500
Less than 10		Less than 25

AZERBAIJAN
GEORGIA
ARM.
TURKMEN.
TURKEY
SYRIA
LEB.
ISRAEL
JORDAN
IRAQ
KAZAKHSTAN
UZBEK.
KYRG.
TAJ.
IRAN
SAUDI ARABIA
KUWAIT
BAHRAIN
QATAR
U.A.E.
AFGHAN.
YEMEN
OMAN
PAKISTAN
NEPAL
Socotra
Yemen
INDIA
BANGLADESH
BHUTA
MALDIVES
SRI LANKA

Tsunami, 1900–2015
Run-up height

Meters		Feet
More than 10		More than 32
5–10	○	16–32

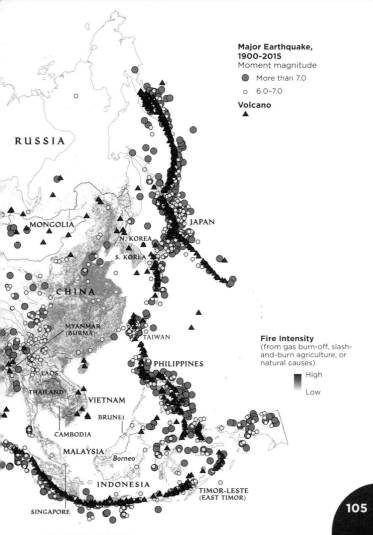

Major Earthquake, 1900-2015
Moment magnitude
- ● More than 7.0
- ○ 6.0–7.0

Volcano
▲

Fire Intensity
(from gas burn-off, slash-and-burn agriculture, or natural causes)

High

Low

RUSSIA

MONGOLIA

N. KOREA

S. KOREA

CHINA

JAPAN

MYANMAR
(BURMA)

TAIWAN

LAOS

PHILIPPINES

THAILAND

VIETNAM

BRUNEI

CAMBODIA

MALAYSIA

Borneo

INDONESIA

TIMOR-LESTE
(EAST TIMOR)

SINGAPORE

Land Cover

- ■ Evergreen needleleaf forest
- ■ Evergreen broadleaf forest
- ■ Deciduous needleleaf forest
- ■ Deciduous broadleaf forest
- ■ Mixed forest
- ■ Woody savanna
- ■ Savanna
- ■ Closed shrubland
- □ Open shrubland
- □ Grassland
- □ Cropland
- □ Barren or sparsely vegetated
- ■ Urban or built-up
- □ Snow and ice
- □ Cropland/natural vegetation mosaic
- ■ Wetland
- ○ Urban area with more than 5 million inhabitants

AZERBAIJAN
GEORGIA
TURKEY
SYRIA
LEB.
ISRAEL
ARM.
KAZAKHSTAN
IRAQ
Baghdad
TURKMEN.
UZBEK.
JORDAN
Tehran
IRAN
KYRG.
KUWAIT
SAUDI
ARABIA
AFGHANISTAN
TAJ.
BAHRAIN
QATAR
U.A.E.
Lahore
PAKISTAN
BHUTAN
YEMEN
OMAN
Delhi
Karachi
NEPAL
Ahmadabad
Dhaka
Surat
INDIA
Kolkata
Mumbai
(Calcutta)
(Bombay)
Pune
BANGLADESH
Socotra
Yemen
Hyderabad
Bengaluru
(Bangalore)
BANGLADESH
Chennai
(Madras)
MALDIVES
SRI
LANKA

RUSSIA

MONGOLIA

Harbin

JAPAN

Shenyang N. KOREA Tokyo
Nagoya
Beijing Tianjin Osaka-Kobe

S. KOREA

CHINA

Xi'an Nanjing Shanghai

Wuhan

Chengdu Chongqing

Dongguan TAIWAN

Guangzhou Shenzhen
Hong Kong

MYANMAR
(BURMA) PHILIPPINES

LAOS

Manila

THAILAND VIETNAM

Bangkok CAMBODIA

Ho Chi Minh City
(Saigon)

New
Guinea

BRUNEI

Kuala Lumpur MALAYSIA

Singapore Borneo

I N D O N E S I A

Sumatra

Jakarta TIMOR-LESTE
(EAST TIMOR)

SINGAPORE Java

Climate Zones

(based on modified Köppen system)

Humid equatorial climate (A)
- ■ No dry season (Af)
- ■ Short dry season (Am)
- ■ Dry winter (Aw)

Dry climate (B)
- ■ Semiarid (BS)   h = hot
- ■ Arid (BW)   k = cold

Humid temperate climate (C)
- ■ No dry season (Cf)
- ■ Dry winter (Cw)   a = hot summer
- ■ Dry summer (Cs)   b = cool summer
                              c = short, cool summer
Humid cold climate (D)   d = very cold winter
- ■ No dry season (Df)
- ■ Dry winter (Dw)

Cold climate (E)
- ■ Tundra and ice

Highland climate (H)
- ■ Unclassified highlands

JORDAN
ISRAEL
AZERBAIJAN
GEORGIA
ARMENIA
TURKM.
UZB.
TURKEY
LEB.
SYRIA
IRAQ
KUWAIT
SAUDI
ARABIA
BAHRAIN
QATAR
U.A.E.
YEMEN
OMAN
IRAN
AFGHAN.
PAK.
KAZAKHSTAN
KYRG.
TAJ.
RUSSIA
MONGOLIA
CHINA
BHUTAN
NEPAL
INDIA
BANGLADESH
MYANMAR
(BURMA)
LAOS
THAILAND
CAMBODIA
VIETNAM
NORTH
KOREA
SOUTH
KOREA
JAPAN
TAIWAN
PHILIPPINES
BRUNEI
MALAYSIA
SRI
LANKA
MALDIVES
INDONESIA
SINGAPORE
TIMOR-LESTE
(EAST TIMOR)

Water Availability

(in millimeters per person
per year)

- ■ More than 750
- ▨ 251–750
- ▨ 26–250
- □ Less than 26
- □ No data

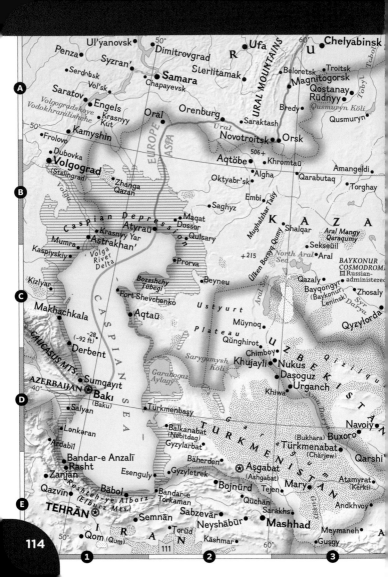

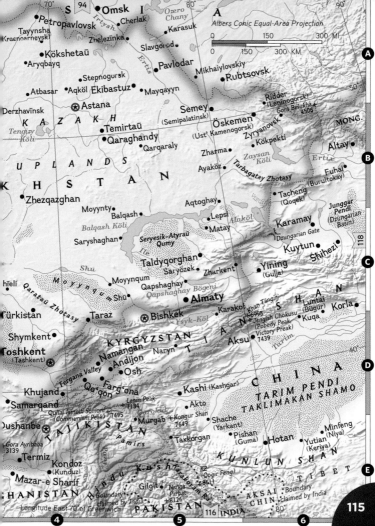

Albers Conic Equal-Area Projection

Mashhad
Kāshmar
114 TURKM.
Mazar-e Sharīf
Kondoz
60°
70°
TAJ KUSH
Nowshak
K2
Kar akoram Rg.
Esfahan
(Isfahan)
Herāt
Bīrjand
Kābol
(Kabul)
Jalālābād
Peshawar
Rawalpindi
Islāmābād
Srinaga
KASH
A
111
Yazd
AFGHANISTAN
Boundary claimed
by India
KASH

I R A N
Dasht-e Kavir
Farah
Qalāt
(Kalat)
Kandahār
Chaman
Islāmābād
Faisalābād
Gujrānwāl
30°
Shīrāz
Kermān
Zābol
Zāhedān
Lashkar Gah
Zaranj
Rudbār
Quetta
LAHORE
Ludhiāna
Amritsa
Chandigarh

Bam
Mīrjāveh
Kūh-e Taftān
4042
Rīgestan
Multan
Bahāwalpur
Thar Desert
(Great Indian Desert)
DELHI

Bandar
'Abbās
Strait of Hormuz
Īrānshahr
Nok Kundi
Bīkaner
Jaipur
New
Delhi
Agra
Gwalior

B
PERSIAN
GULF
Khasab
Turbat
Pasni
Larkāna
Sukkur
Jodhpur
Ajmer
Kota

Dubai
Gwadar
Sonmiani Bay
Hyderābād
Udaipur
Bhopal

Abū Zaby
(Abu Dhabi)
As Sīb
Matrah
KARACHI
Rann of Kutch
AHMADĀBĀD
(Ahmedabad)
Ujjain
Indore

U.A.E.
Jabal ash Shām
2980
Masqat
(Muscat)
TROPIC OF CANCER
Rajkot
Gulf of Kutch
Vadodara
Surat
Amravati

113
O M A N
Ra's al Hadd
Porbandar
Gulf of Khambhat
Aurangābād
Jalna

C
20°
Jazīrat Masīrah
Thane
(Bombay) MUMBAI
Kalyan
Nanded

Ra's ash Sharbatāt
Jazā'ir al Hallānīyāt
(Kuria Muria Is.)
A R A B I A N
PUNE
Sholapur

S E A
Kolhapur
Belagavi
(Belgaum)
Marmagao,
Kurnoo
Ballari
(Bellary)

D
Azimuthal Equidistant Projection
0   250   500 MI
0   250   500 KM
(Hubli) Hubballi
Davangere
(Bangalore) BENGALURU

10°
Lakshadweep
India
(Mangalore) Mangaluru
(Mysore) Mysuru
Coimbatore
Thrissur
(Trichur)

(Cochin) Kochi
Madurai
Tirunelveli
Nine Degree Channel
Thiruvananthapuram
(Trivandrum)
Eight Degree Channel
Cape Comorin

E
I N D I A N
MALDIVES

116
60°
Longitude East 70° of Greenwich

1   2   3

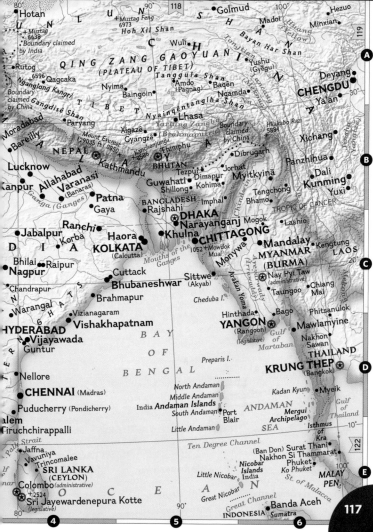

Hotan

Muztag 6639
Boundary claimed by India

KUN LUN SHAN

Golmud

Muztag Feng 6973
Hoh Xil Shan

Hezuo

Madoi

Minxian

Huang Yellow

QING ZANG GAOYUAN
(PLATEAU OF TIBET)

Rutog

Nganglong Kangri 6596
Boundary claimed by China
Gangdise Shan

Qagcaka

Nyima

Baingoin

Wuli

Bayan Har Shan

Yushu (Gyêgu)

Amdo (Pagnag)

Baqên

Ngamda

CHENGDU

Deyang

Ya'an

T I B E T

Tanggula Shan

Moradabad

Bareilly

Paryang

Xigaze

Gyangzê

Nyainqêntanglha Shan

Lhasa

Yarlung Zangbo

Brahmaputra

Boundary claimed by China

Hkakabo Razi 5881

Xichang

Panzhihua

Dali

Kunming

Yuxi

Mount Everest (29035 ft) 8850

Kanchenjunga 8586

Thimphu

Dibrugarh

Lucknow

Allahabad

Varanasi (Banaras)

NEPAL

Kathmandu

BHUTAN

Tezpur

Jorhat

Myitkyina

anpur

Ganga (Ganges)

Patna

Gaya

Guwahati

Shillong

Dimapur

Kohima

Imphal

Tengchong

Bhamo

Lashio

Ranchi

Korba

Jabalpur

BANGLADESH

Rajshahi

DHAKA

Narayanganj

Mogok

Mandalay

Kengtung

LAOS

Bhilai

Nagpur

Raipur

Haora

KOLKATA (Calcutta)

Khulna

CHITTAGONG

1052 Mowdok Mual

Monywa

MYANMAR (BURMA)

Chandrapur

Cuttack

Bhubaneshwar

Mouths of the Ganges

Sittwe (Akyab)

Nay Pyi Taw (administrative)

Taungoo

Chiang Mai

Warangal

Vizianagaram

Brahmapur

Cheduba I.

Arakan Yoma

Hinthada

Bago

Phitsanulok

Nakhon Sawan

HYDERABAD

Vijayawada

Guntur

Vishakhapatnam

BAY

OF

BENGAL

Preparis I.

YANGON (Rangoon) (legislative)

Gulf of Martaban

Mawlamyine

THAILAND

KRUNG THEP (Bangkok)

Nellore

CHENNAI (Madras)

Puducherry (Pondicherry)

North Andaman

Middle Andaman

India Andaman Islands

South Andaman

Port Blair

Little Andaman

Kadan Kyun

ANDAMAN

Mergui Archipelago

Myeik

Gulf of Thailand

Isthmus of Kra

alem

iruchchirappalli

Little Andaman

SEA

Palk Strait

Ten Degree Channel

Jaffna

Vavuniya

Trincomalee

SRI LANKA (CEYLON)

Colombo (administrative)

2524

Sri Jayewardenepura Kotte (legislative)

O C E A N

(Ban Don) Surat Thani

Nakhon Si Thammarat

Phuket

Ko Phuket

MALAY PEN.

Nicobar Islands India

Little Nicobar

Great Nicobar

INDONESIA

Sumatra

Banda Aceh

Great Channel

St. of Malacca

TROPIC OF CANCER

117

Longitude East 130° of Greenwich

Siping, Liaoyuan, Panshi, Huadian, Yanji, Tumen, Vladivostok
RUSSIA
Dongfeng, Meihekou, Hunan, Helong, Unggi (Suonbang), Mys Gamova
Tieling, CHINA, Liuhe, Songjianghe, Baishan, Musan, Rajin (Najin)
SHENYANG, Fushun, +Paektu-san 2744, Nanam, Ch'ŏngjin
Liaoyang, Tonghua, Chasong, Hyesan, Kyŏngsŏng
Benxi, Manp'o, Kimhyŏnggwŏn, Kilju
Anshan, Kanggye, Musudan
Fengcheng, Sakchu, Ch'osan, Kimch'aek
Dandong, Hŭich'ŏn, Pukch'ŏng, Tanch'ŏn
Zhuanghe, Sinŭiju, Yŏngbyŏn, KOREA, Hamhŭng
Sŏnch'ŏn, Anju, Tŏkch'ŏn, Hŭngnam
Sŏjosŏn-man, P'yŏng-sŏng, Sunch'ŏn, Okp'yŏng, Yŏnghŭng-man
KOREA BAY, P'yŏngyang, Wŏnsan, Anbyŏn
Namp'o, Songnim
Anak, Sariwŏn, Sep'o, Albers Conic Equal-Area Projection
Changyŏn, Ich'ŏn, P'yŏnggang, Military Demarcation Line, July 27, 1953
Changsan-got, Haeju, Gimhwa, Sokcho, SEA OF JAPAN (EAST SEA)
Baengnyeongdo, Ongjin, Kaesŏng, Chuncheon
Northern Limit Line July 27, 1953, Goyang, SEOUL, Gangneung
Incheon (Inch'ŏn), Seongnam, Donghae, Samcheok, Ulleungdo (Dagelet) S. Korea
Ansan, Wonju, Jecheon
YELLOW SEA, Suwon
Anseong, Cheonan
The Democratic People's Republic of Korea is referred to as North Korea. The Republic of Korea is known as South Korea.
Seosan, Cheongju, Andong, Hupo
SOUTH KOREA
(Taejŏn) Daejeon, Uiseong
Ganggyeong, Gumi, Pohang
Gunsan, Iksan, Daegu, Gyeongju (Kyŏngju)
Jeonju, (Taegu), Ulsan
Namwon, Changwon
(Kwangju) Gwangju, +1915, Busan (Pusan)
Gochang, Jinju, Goseong, Masan
Hampyeong, Suncheon, Yeosu
Mokpo, Gangjin, Boseong, Geojedo, Korea Strait
Jindo, Haenam, Kamino Shima, Kawashiri Misaki
Soheuksando S. Korea, Wando, Tsushima Japan, HONSHŪ
Shimono Shima, Shimonoseki, Suō Nada
Jeju Strait, Kitakyūshū
(Cheju) Jeju, Jeju-Do S. Korea, JAPAN
Hallim, Hallasan 1950, Seogwipo, Fukuoka, KYŪSHŪ
Saseho, Saga

0   50   100 MI
0   50   100 KM

Yichun
Hailun
Suihua
Hegang
Heilongjiang
Jiamusi
Hulin
Bikin
Khor
Ostrov
Sakhalin
Yuzhno
Sakhalinsk
SEA OF
OKHOTSK
CHINA
Jixi
Lesozavodsk
Amur
Svetlaya
Mys Kril'on
La Perouse Strait
Sōya Misaki
Mys Aniva
Zaliv
Aniva
A
Mudanjiang
Dal'nerechensk
Amgu
Rubun Tō
Rishiri Tō
Wakkanai
Kunashiri
(Kunashir)
Russia
Dunhua
Ussuriysk
Spassk Dal'niy
Terney
Nayoro
Asahikawa
+2290
Kitami
HOKKAIDŌ
Kushiro
Ozero Khanka
(Xingkai Hu)
Vladivostok
Nakhodka
Ishikari Wan
Kamui Misaki
Otaru
Sapporo
Tomakomai
Obihiro
Erimo Misaki
NORTH
KOREA
Ch'ŏngjin
Zaliv
Petra
Velikogo
(Peter the
Great Bay)
Okushiri Tō
Hakodate
Uchiura Wan
B
Shirakami Misaki
Tsugaru Kaikyō
Mutsu Wan
Aomori
Z
SEA OF JAPAN
(EAST SEA)
Hirosaki
Hachinohe
Morioka
Kamaishi
SOUTH
KOREA
Military Demarcation Line,
July 27, 1953
Nyūdō Zaki
Akita
Tobi Shima
Sakata
Yamagata
Ishinomaki
Sendai
Ishinomaki Wan
C
Daegu
(Taegu)
Ulleungdo
(Dagelet)
S. Korea
Dokdo (Takeshima,
Liancourt Rocks)
S. Korea
Niigata
Sado
Toyama
Maebashi
Nagano
Fukushima
Utsunomiya
Mito
Oki Shotō
Tottori
Kanazawa
Matsumoto
N
Kōfu
SAITAMA
TŌKYŌ
Kawasaki
Matsue
Miho Wan
Hinomi Saki
Kyōto
Nagoya
+Fuji
3776
Yokohama
Korea Strait
Hiroshima
Takamatsu
KŌBE
ŌSAKA
Hamamatsu
Shizuoka
Tōkyō Wan
PACIFIC
OCEAN
D
Yamaguchi
Tsushima
Sūrizit
Kitakyūshū
SHIKOKU
Tokushima
Wakayama
Toyohashi
Ise Wan
Sagami Nada
Fukuoka
Nagasaki
Ōita
Matsuyama
Kōchi
Shiono Misaki
Kumano Nada
Gotō
Rettō
Kumamoto
Bungo Suidō
J
KYŪSHŪ
Miyazaki
Hachijō Jima
Noma
Misaki
Kagoshima
Kanoya
Ōsumi Kaikyō
(Van Diemen Strait)
PHILIPPINE SEA
Aoga Shima
Beyonesu Retsugan
Yaku Shima
Tanega Shima
Sumisu Jima
(Smith)
Tokara Kaikyō
(Colnett Strait)
Albers Conic Equal-Area Projection
Tori Shima
E
Tokara Rettō
Nakano Shima
0   100   200 MI
0   100   200 KM
Sōfu Gan
(Lot's Wife)

TAIWAN
120°
130°
140°
20°

Kaohsiung
Bashi Channel
Batan Islands
PACIFIC OCEAN

LUZON STRAIT
A

Babuyan Islands

aoag
Aparri
Tuguegarao
igan
Mount Pulog 2934
ando
LUZON
aguio
Mt. Pinatubo 1780
geles
Quezon City
atangas
MANILA

Catanduanes
Legazpi
Sorsogon
San Bernardino Str.
B

Mindoro
Masbate
Calbayog
Samar
amian up
Masbate
Tacloban
Leyte

FEDERATED STATES
OF MICRONESIA
160°

Roxas
Panay
Cebu
Iloilo
Bacolod
Cebu
Bohol
Ulithi Atoll
10°
Fais

Negros
Bohol Sea
Slargao
Yap Islands

uerto
rincesa
Pagadian
Butuan
Cagayan de Oro
Iligan
MINDANAO
PALAU
Melekeok
Babelthuap
Ngulu Atoll
Sorol Atoll
CAROLINE ISLANDS

SULU SEA
Mount Apo
2954
mboanga
Davao Gulf
Davao

Basilan
Moro
Gulf
General Santos
C

Sonsorol Islands

Tawi Tawi
Sulu Arch.
Pulo Anna
Merir

vel Bay

Kepulauan
Talaud
Tobi
Helen Island

CELEBES SEA
Kep. Sangihe

Manado
Morotai

Tolitoli
1635
Halmahera
EQUATOR
0°

Gorontalo
Ternate

Teluk
Tomine
Bacan
M
Halmahera
Sea
Waigeo
Supiori
Biak
Tanjung D'Urville

lu
N
Peleng
Mangole
Obi
Sorong
Yapen
Sarmi

SULAWESI
(CELEBES)
Taliabu
Kep. Sula
Salawati
Batanme
A
Teluk
Cenderawasih
Jayapura
Aitape
D

478
Teluk Tolo
Buru
CERAM
Seram (Ceram)
Selat Manipa
Fakfak
Puncak Jaya
4884
NEW GUINEA

S
Parepare
Wawonii
Muna
Ambon
BANDA SEA
Timika
PAPUA

Makassar
(Ujungpandang)
Kendari
Buton
Baubau
Kep. Kai
Tual
Wokam
Kep.
Aru
Teluk
Flamingo
Kepi
NEW
GUINEA

elayar
BANDA SEA
Kobroor
Muting

Kalao
Lombien
Pantar
Kep. Barat Daya
Romang
Trangan
Merauke
E

aba
Ruteng
Flores
Alor
Wetar
Babar
Moa
Yamdena
Dolak
Tanjung Vals

NDA ISLANDS
Dili
Timor
TIMOR-LESTE
(EAST TIMOR)
Selaru
ARAFURA

Waingapu
Kupang
Roti
TIMOR SEA
SEA
10°

120°
Sawu
130°
AUSTRALIA
149
140°

Azimuthal Equidistant Projection
0   250   500 MI
0   250   500 KM

Molucca Sea
Celebes Sea
Philippine Sea
Moro Gulf
Flores Sea
Sawu

123

ALGERIA

Cairo

Niger

NIGERIA

Lagos

Congo

Lake Victoria

Kinshasa

Lake Tanganyika

Luanda

Johannesburg

AFRICA

GEOGRAPHIC EXTREMES

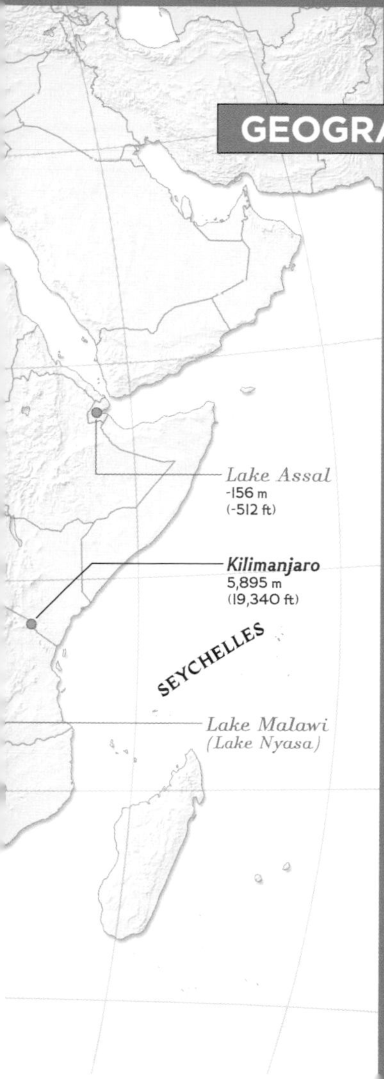

CONTINENTAL POLITICAL FACTS

TOTAL NUMBER OF COUNTRIES: 54

LARGEST COUNTRY BY AREA: Algeria
2,381,741 sq km (919,595 sq mi)

SMALLEST COUNTRY BY AREA: Seychelles   455 sq km (176 sq mi)

MOST POPULOUS COUNTRY: Nigeria
181,562,000

LEAST POPULOUS COUNTRY: Seychelles
92,000

LARGEST URBAN AREAS BY POPULATION:
Cairo, Egypt 18,722,000
Kinshasa, Dem. Rep. Congo
11,587,000
Lagos, Nigeria   10,580,000
Johannesburg, South Africa
9,399,000
Luanda, Angola   5,506,000

CONTINENTAL PHYSICAL FACTS

AREA: 30,065,000 sq km
(11,608,000 sq mi)

HIGHEST POINT: Kilimanjaro,
Tanzania   5,895 m (19,340 ft)

LOWEST POINT: Lake Assal, Djibouti
-156 m (-512 ft)

LONGEST RIVERS:
Nile   6,700 km (4,160 mi)
Congo   4,700 km (2,900 mi)
Niger   4,170 km (2,590 mi)

LARGEST NATURAL LAKES:
Lake Victoria
69,500 sq km (26,800 sq mi)
Lake Tanganyika
32,600 sq km (12,600 sq mi)
Lake Malawi (Lake Nyasa)
28,900 sq km (11,200 sq mi)

Lake Assal
-156 m
(-512 ft)

Kilimanjaro
5,895 m
(19,340 ft)

SEYCHELLES

Lake Malawi
(Lake Nyasa)

125

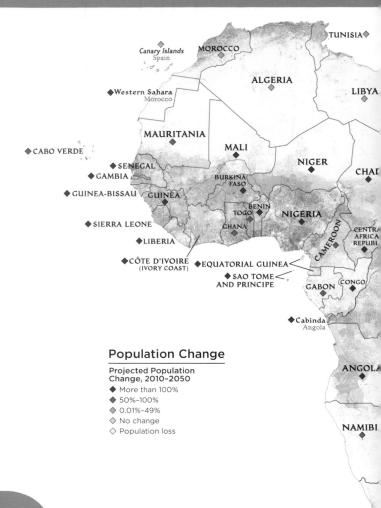

Population Change

Projected Population
Change, 2010-2050

◆ More than 100%
◆ 50%-100%
◇ 0.01%-49%
◇ No change
◇ Population loss

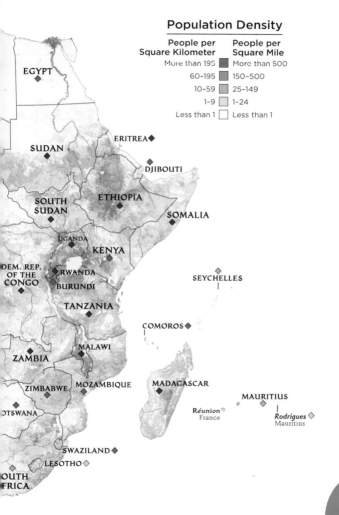

Population Density

People per Square Kilometer	People per Square Mile
More than 195	More than 500
60–195	150–500
10–59	25–149
1–9	1–24
Less than 1	Less than 1

EGYPT

ERITREA

SUDAN

DJIBOUTI

SOUTH SUDAN

ETHIOPIA

SOMALIA

UGANDA

KENYA

DEM. REP. OF THE CONGO

RWANDA

BURUNDI

SEYCHELLES

TANZANIA

COMOROS

MALAWI

ZAMBIA

ZIMBABWE

MOZAMBIQUE

MADAGASCAR

MAURITIUS

Réunion
France

Rodrigues
Mauritius

OTSWANA

SWAZILAND

LESOTHO

OUTH FRICA

Fire Intensity
(from gas burn-off, slash-and-burn agriculture, or natural causes)

High

Low

TUNISIA

MOROCCO

Canary Islands
Spain

ALGERIA

LIBYA

Western Sahara
Morocco

MAURITANIA

MALI

NIGER

CHA

CABO
VERDE

SENEGAL

GAMBIA

GUINEA-BISSAU

GUINEA

BURKINA
FASO

NIGERIA

SIERRA LEONE

GHANA

TOGO

BENIN

CAMEROON

LIBERIA

CÔTE D'IVOIRE
(IVORY COAST)

EQUATORIAL GUINEA

SAO TOME
AND PRINCIPE

CONGO

GABON

Cabinda
Angola

**Major Earthquake,
1900-2015**
Moment magnitude

More than 7.0

6.0–7.0

Volcano

ANGOL

NAMIB

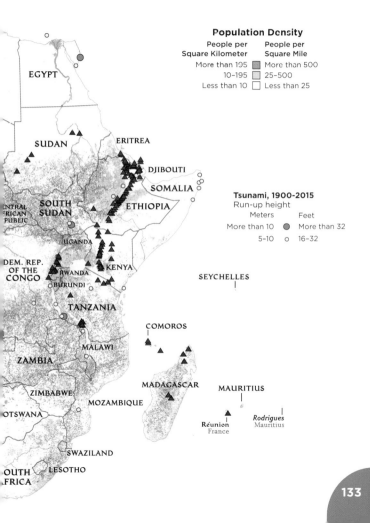

Population Density

People per Square Kilometer		People per Square Mile
More than 195		More than 500
10–195		25–500
Less than 10		Less than 25

EGYPT

SUDAN

ERITREA

DJIBOUTI

SOMALIA

Tsunami, 1900-2015
Run-up height

Meters	Feet
More than 10	More than 32
5–10	16–32

SOUTH SUDAN

ETHIOPIA

CENTRAL AFRICAN REPUBLIC

DEM. REP. OF THE CONGO

UGANDA

RWANDA

KENYA

SEYCHELLES

BURUNDI

TANZANIA

COMOROS

MALAWI

ZAMBIA

MADAGASCAR

MAURITIUS

ZIMBABWE

MOZAMBIQUE

BOTSWANA

Rodrigues
Mauritius

Réunion
France

SWAZILAND

SOUTH AFRICA

LESOTHO

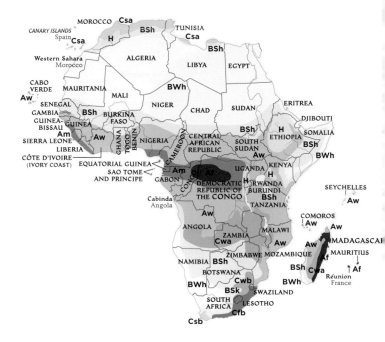

Climate Zones

(based on modified Köppen system)

Humid equatorial climate (A)
- No dry season (Af)
- Short dry season (Am)
- Dry winter (Aw)

Dry climate (B)
- Semiarid (BS) } h = hot
- Arid (BW) } k = cold

Humid temperate climate (C)
- No dry season (Cf)
- Dry winter (Cw) } a = hot summer
- Dry summer (Cs) } b = cool summer

Highland climate (H)
- Unclassified highlands

Water Availability

(millimeters per person
per year)

	More than 750
	251–750
	26–250
	Less than 26

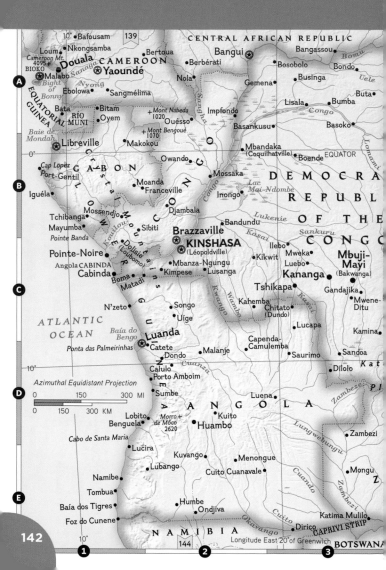

10° •Bafousam
Loum• •Nkongsamba •Bertoua CENTRAL AFRICAN REPUBLIC •Bangassou
Cameroon Mt. **Douala** CAMEROON •Berbérati •Bangui •Bosobolo •Bondo Bomu
4095• •Yaoundé •Nola •Gemena •Businga •Buta
BIOKO •Malabo Sanaga Nyong •Ebolowa •Sangmélima •Impfondo •Lisala •Bumba Uele
Bight •Bata RÍO •Bitam +Mont Nabeda •Basankusu •Basoko Congo
of MUNI •Oyem 1020 •Ouésso
Bonny EQUATORIAL +Mont Bengoué
GUINEA Baie de 1070• •Makokou •Mbandaka EQUATOR
Mondah •Libreville 0° •Owando (Coquilhatville) •Boende Lomami
Cap Lopez GABON •Mossaka DEMOCRA
•Port-Gentil •Moanda Lac REPUBL
Iguéla• •Franceville Inongo• Mai-Ndombe OF THE
Tchibanga• •Mossendjo Lukenie CONGO
Mayumba• •Sibiti •Bandundu Sankuru Mbuji-
Pointe Banda Dolisie Djambala •Brazzaville Kasai •Ilebo •Mweka Mayi
Pointe-Noire (Loubomo) ⊛**KINSHASA** •Kikwit •Luebo **Kananga** (Bakwanga)
Angola CABINDA •Mbanza-Ngungu (Léopoldville) Lusanga •Tshikapa •Gandajika
Cabinda• •Kimpese Kwango •Kahemba •Chitato •Mwene-
Boma• •Matadi Wamba (Dundo) Ditu
N'zeto• •Songo •Lucapa •Kamina
ATLANTIC •Uíge •Capenda-
OCEAN Baía do ⊛**Luanda** Camulemba •Saurimo •Sandoa
Bengo •Catete •Malanje •Dilolo Kat
Ponta das Palmeirinhas •Dondo Cuanza
10° •Calulo
Azimuthal Equidistant Projection •Porto Amboim Zambeze
0 150 300 MI •Sumbe •Luena **ANGOLA** Lungwebungu
0 150 300 KM •Lobito Morro+ •Kuito •Zambezi
•Benguela de Móco •Huambo •Mongu Z
Cabo de Santa Maria 2620 •Lucira •Kuvango •Menongue Zambezi
•Namibe •Lubango •Cuito Cuanavale •Cuando
•Tombua Okavango •Dirico CAPRIVI STRIP
Baía dos Tigres• •Humbe •Katima Mulilo
Foz do Cunene• •Ondjiva Cuito Longitude East 20° of Greenwich BOTSWANA
10° NAMIBIA 144

1 2 3

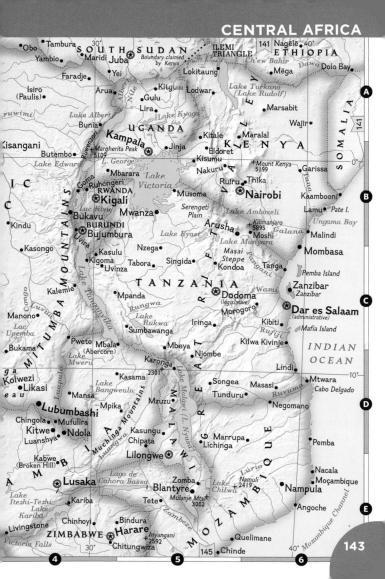

Obo •Tambura• 30°
•Yambio
•Maridi
SOUTH SUDAN
☆Juba
ILEMI
TRIANGLE
Boundary claimed
by Kenya
Nagēlē• 40°
Dolo Bay•
141
ETHIOPIA
Ch'ew Bahir
Dawa
Mēga

•Isiro
(Paulis)
•Faradje
•Yei
Lokitaung
Lake Turkana
(Lake Rudolf)
•Marsabit

•Arua
Albert
Nile
•Gulu
•Kitgum
•Lira
•Lodwar
Wajīr•
A

~ruwimi
•Bunia
Lake Albert
Lake Kyoga
Maralal•

•Kisangani
•Butembo
UGANDA
Ruwenzori
Margherita Peak
5109
☆Kampala
•Jinja
•Kitale
•Eldoret
•Kisumu
Nakuru
•Maralal
+ Mount Kenya
5199
K E N Y A
•Garissa
Kaambooni•
0°

Lake Edward
L. George
Goma
Ruhengeri•
RWANDA
☆Kigali
•Mbarara
•Musoma
Rutru
Nairobi
Thika
Lamu• •Pate I.
B
Ungama Bay

•Kindu
Bukavu•
BURUNDI
☆Bujumbura
Lake
Victoria
Serengeti
Plain
•Mwanza
Arusha
Lake Amboseli
Kilimanjaro
+ 5895
Moshi
Malindi•
•Mombasa

•Kasongo
Uvira
Kasulu•
Kigoma•
•Nzega
Lake Eyasi
Masai
Steppe
Lake Manyara
Kondoa
Tanga•
Pemba Island
•Zanzibar

Lake Tanganyika
•Uvinza
•Tabora
•Singida
T A N Z A N I A
Pangani

•Manono
Luvua
Lac
Upemba
•Kalemie
Mpanda•
Bungwa
☆Dodoma
(legislative)
•Morogoro
Wami
Zanzibar
Dar es Salaam
(administrative)
C

•Bukama
Congo
MITUMBA MOUNTAINS
Pweto
(Abercorn)
Lake
Rukwa
Sumbawanga•
•Iringa
Rufiji
•Kibiti
Kilwa Kivinje
Mafia Island
INDIAN
OCEAN

Kolwezi•
•Likasi
Mbala
(Abercorn)
Lake
Mweru
•Kasama
Karonga
2301
•Mbeya
•Njombe
•Lindi
10°

ga
eau
•Lubumbashi
•Mansa
Lake
Bangweulu
•Mpika
Mzuzu•
Songea•
•Tunduru
•Masasi
Ruvuma
Mtwara•
Cabo Delgado
Negomano•
D

Chingola•
Kitwe•
Luanshya•
•Mufulira
•Ndola
Z
Muchinga Mountains
Luangwa
•Kasungu
•Chipata
Lichinga•
•Marrupa
Pemba•

•Kabwe
(Broken Hill)
M
B
I
A
Lilongwe
MALAWI
L. Nyasa
Lúrio
MOZAMBIQUE
•Nacala
•Moçambique
Nampula•

A
☆Lusaka
•Kariba
Lago de
Cahora Bassa
Zomba•
Blantyre•
+
Lake
Chilwa
Namuli
+ 2419
Mozambique Channel
E

Lake
Itezhi-Tezhi
Lake
Kariba
•Chinhoyi
•Bindura
•Tete
Mulanje Mts.
3002
Quelimane•

•Livingstone
Victoria Falls
ZIMBABWE
☆Harare
Zambeze
Inyangani
+ 2592
•Chitungwiza
30°
145 •Chinde
40°
Angoche•

4
5
6

A U S T R A L I A

Lake Eyre
-16 m (-52 ft)

Brisbane

Lake Gairdner

Lake Torrens

Darling

Perth

Adelaide

Murray

Melbourne

Murrumbidgee

Mount Kosciuszko
2,228 m
(7,310 ft)

Sydney

NAURU

AUSTRALIA & OCEANIA

GEOGRAPHIC EXTREMES

CONTINENTAL POLITICAL FACTS

TOTAL NUMBER OF COUNTRIES: 14

LARGEST COUNTRY BY AREA: Australia
7,741,220 sq km (2,988,901 sq mi)

SMALLEST COUNTRY BY AREA:
Nauru 21 sq km (8 sq mi)

MOST POPULOUS COUNTRY: Australia
22,993,000

LEAST POPULOUS COUNTRY: Nauru
9,500

LARGEST URBAN AREAS BY POPULATION:
Sydney, Australia 4,505,000
Melbourne, Australia 4,203,000
Brisbane, Australia 2,202,000
Perth, Australia 1,861,000
Auckland, New Zealand 1,344,000

CONTINENTAL PHYSICAL FACTS

AREA: 7,741,220 sq km (2,988,901 sq mi)

HIGHEST POINT: Mount Kosciuszko,
New South Wales, Australia
2,228 m (7,310 ft)

LOWEST POINT: Lake Eyre, South
Australia, Australia -16 m (-52 ft)

LONGEST RIVERS:
Murray 2,510 km (1,560 mi)
Darling 1,550 km (960 mi)
Murrumbidgee 1,490 km (930 mi)

LARGEST NATURAL LAKES:
Lake Eyre 0–9,700 sq km
(0–3,700 sq mi)
Lake Torrens 0–5,700 sq km
(0–2,200 sq mi)
Lake Gairdner 0–4,300 sq km
(0–1,680 sq mi)

Auckland

147

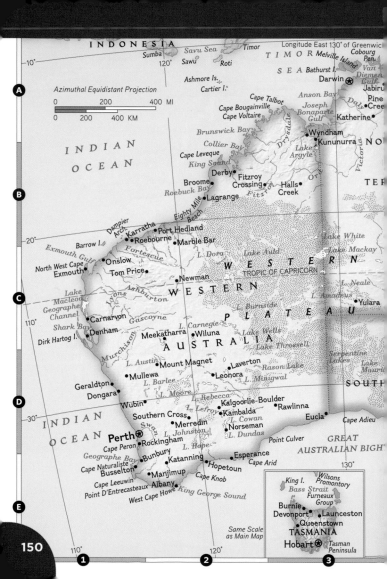

INDONESIA

Sumba   Savu Sea   Timor   Longitude East 130° of Greenwic

120°   Sawu   Roti   TIMOR   Melville Island   Cobourg
                                  Pen.
10°   Van
        Ashmore Is.   S E A   Darwin ⊛   Diemen
        Cartier I.   Gulf
                    Cape Talbot   Anson Bay   Jabiru
                Cape Bougainville   Joseph   Pine
                Cape Voltaire   Bonaparte   Da   Cree
                              Gulf   Katherine
        Brunswick Bay   Lake   Wyndham   NO
        Collier Bay   Argyle   Kununurra
Cape Leveque   Derby   Fitzroy   TER
INDIAN   King Sound   Crossing   Halls
OCEAN   Broome   Fitzroy   Creek
        Roebuck Bay   Lagrange
                    Eighty Mile Beach

        Dampier   Lake White
        Arch.   Lake Mackay
20°   Barrow I.   Karratha   Port Hedland
Exmouth Gulf   Roebourne   Marble Bar
North West Cape   Onslow   Fortescue   L. Dora   WESTERN
Exmouth   Tom Price   Newman   TROPIC OF CAPRICORN
                        Lake Auld

Lake   Ashburton   WESTERN   L. Neale
Macleod
Channel   Carnarvon   Gascoyne   PLATEAU   L. Amadeus
110°   Shark Bay   L. Burnside   Yulara
Dirk Hartog I.   Denham   Meekatharra   Wiluna   L. Carnegie   Lake Wells
                                    Lake Throssell
        Murchison   AUSTRALIA   Serpentine
        L. Austin   Mount Magnet   Lakes
Geraldton   Mullewa   Laverton   Rason Lake   Lake
        L. Barlee   Leonora   Mauri
Dongara   L. Moore   L. Rebecca   Minigwal   SOUTH
        Wubin   L. Lefroy   Kalgoorlie-Boulder   Rawlinna
INDIAN   Southern Cross   Kambalda
30°   OCEAN   Merredin   L. Cowan   Eucla   Cape Adieu
        Perth ⊛   Rockingham   Norseman
        Cape Peron   L. Johnston   L. Dundas   Point Culver   GREAT
        Geographe Bay   Bunbury   L. Hope   Esperance   AUSTRALIAN BIGH
        Cape Naturaliste   Katanning   Cape Arid   130°
        Busselton   Manjimup   Hopetoun
        Cape Leeuwin   Albany   Cape Knob
        Point D'Entrecasteaux   King George Sound   King I.   Wilsons
        West Cape Howe   Bass Strait   Promontory
                                Furneaux
                        Same Scale   Burnie   Group
                        as Main Map   Devonport   Launceston
                                Queenstown
                                TASMANIA
                            Hobart ⊛   Tasman
                                    Peninsula

150

Azimuthal Equidistant Projection
0   200   400 MI
0   200   400 KM

110°   120°   130°
1   2   3

Torres Strait
Cape York
Great Barrier Reef
Temple Bay
PAPUA NEW GUINEA

Wessel Is.
Cape Wilberforce
Endeavour Strait
Gov. Pen.
Nhulunbuy
Albatross Bay
Cape Weymouth
PACIFIC OCEAN

Alyangula
Groote Eylandt
GULF OF CARPENTARIA
Weipa
Aurukun
Coen
Princess Charlotte Bay

Ngukurt
Limmen Bight
HERN
Borroloola
Mornington Island
Cape Flattery
Cooktown
Cape Tribulation

Newcastle Waters
Lake Woods
ITORY
Burketown
Normanton
Georgetown
Croydon
Cairns
Innisfail
Rockingham Bay
Hinchinbrook I.
Magdelaine Cays
CORAL SEA ISLANDS
Australia

Lake Sylvester
ennant Creek
Camooweal
Mount Isa
Cloncurry
Hughenden
Charters Towers
Townsville
Ayr Bowen
Mackay
Great Barrier Reef
Repulse Bay
Swan Reefs

Barrow Creek
QUEENSLAND
Winton
Lake Galilee
Capricorn Chan.

Alice Springs
Birdsville
Longreach
Emerald
Yeppoon
Rockhampton
Cape Manifold
Mount Morgan
Gladstone

Oodnadatta
Quilpie
Charleville
Roma
Bundaberg
Hervey Bay
Maryborough
Gympie
Nambour
Fraser Island (Great Sandy I.)
Double Island Pt.
Caloundra
Bongaree

Lake Eyre North
Marree
Leigh Creek
Lake Frome
Cunnamulla
Toowoomba
Brisbane
Gold Coast
Tweed Heads
Lismore
Cape Byron

USTRALIA
Lake Gairdner
Lake Torrens
Collarenebri
Bourke
Walgett
Moree
Goondiwindi
Grafton
Coffs Harbour

Ceduna
Port Augusta
Whyalla
Broken Hill
NEW SOUTH WALES
Tamworth
Armidale
Port Macquarie
Taree

Port Lincoln
C. Carnot
Port Pirie
Wallaroo
Gawler
Mildura
Dubbo
Orange
Bathurst
Maitland
Newcastle
Broken Bay

Kangaroo I.
Adelaide
Wagga Wagga
Wodonga Albury
Canberra
JERVIS BAY TERR.
Sydney
Wollongong
TASMAN SEA

Mount Gambier
Bendigo
Ballarat
VICTORIA
Shepparton
AUSTRALIAN CAPITAL TERR.

Cape Nelson
Warrnambool
Geelong
Melbourne
Traralgon
Cape Howe
Wilsons Promontory
Bass Str.

151

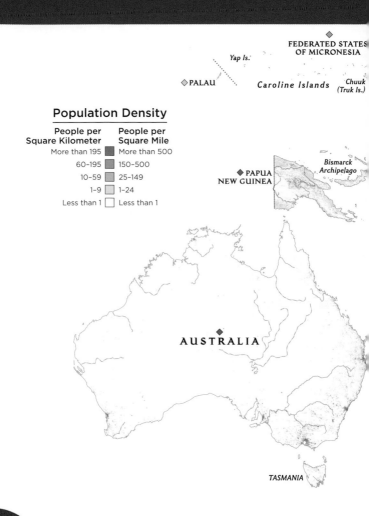

FEDERATED STATES
OF MICRONESIA

Yap Is.

◇ PALAU   Caroline Islands   Chuuk
                                      (Truk Is.)

Population Density

People per Square Kilometer	People per Square Mile
More than 195	More than 500
60–195	150–500
10–59	25–149
1–9	1–24
Less than 1	Less than 1

Bismarck
Archipelago

◆ PAPUA
NEW GUINEA

AUSTRALIA

TASMANIA

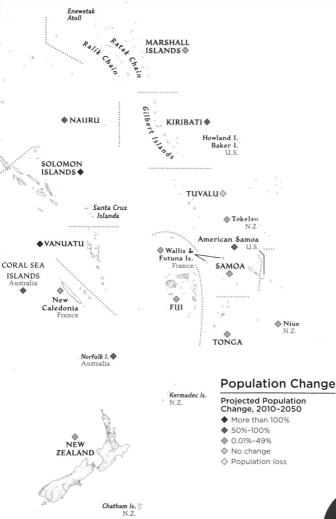

Enewetak
Atoll

Ratak Chain
Ralik Chain

MARSHALL
ISLANDS ◆

NAURU ◆

Gilbert Islands

KIRIBATI ◆

Howland I.
Baker I.
U.S.

SOLOMON
ISLANDS ◆

Santa Cruz
Islands

TUVALU ◇

Tokelau
N.Z. ◆

American Samoa
U.S. ◆

VANUATU ◆

Wallis &
Futuna Is. ◆
France

SAMOA

CORAL SEA
ISLANDS
Australia ◆

New
Caledonia ◆
France

FIJI ◆

Niue
N.Z. ◆

Norfolk I. ◆
Australia

TONGA ◆

Population Change

Kermadec Is.
N.Z.

Projected Population
Change, 2010–2050

◆ More than 100%
◆ 50%–100%
◇ 0.01%–49%
◇ No change
◇ Population loss

NEW
ZEALAND ◆

Chatham Is.
N.Z.

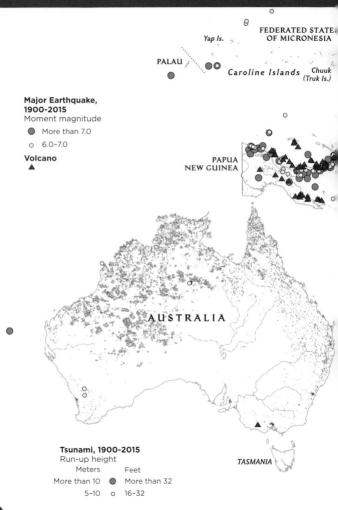

FEDERATED STATE
OF MICRONESIA

Yap Is.

PALAU

Caroline Islands   Chuuk
(Truk Is.)

**Major Earthquake,
1900-2015**
Moment magnitude

⬤ More than 7.0

○ 6.0–7.0

Volcano

▲

PAPUA
NEW GUINEA

AUSTRALIA

Tsunami, 1900-2015
Run-up height

Meters		Feet
More than 10	⬤	More than 32
5–10	○	16–32

TASMANIA

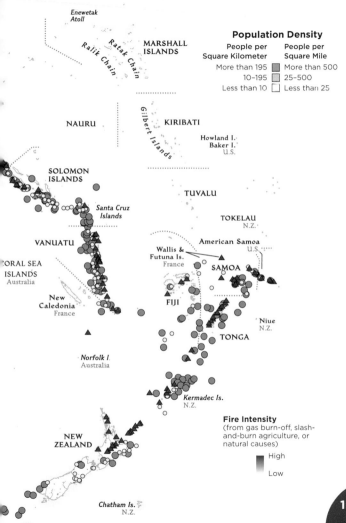

Population Density

People per Square Kilometer	People per Square Mile
More than 195	More than 500
10–195	25–500
Less than 10	Less than 25

Enewetak Atoll

Ralik Chain

Ratak Chain

MARSHALL ISLANDS

NAURU

Gilbert Islands

KIRIBATI

Howland I.
Baker I.
U.S.

SOLOMON ISLANDS

TUVALU

TOKELAU
N.Z.

Santa Cruz Islands

American Samoa
U.S.

Wallis & Futuna Is.
France

SAMOA

VANUATU

CORAL SEA ISLANDS
Australia

FIJI

Niue
N.Z.

New Caledonia
France

TONGA

Norfolk I.
Australia

Kermadec Is.
N.Z.

Fire Intensity
(from gas burn-off, slash-and-burn agriculture, or natural causes)

High

Low

NEW ZEALAND

Chatham Is.
N.Z.

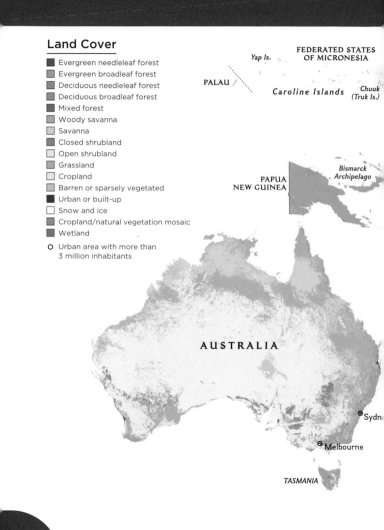

Land Cover

- Evergreen needleleaf forest
- Evergreen broadleaf forest
- Deciduous needleleaf forest
- Deciduous broadleaf forest
- Mixed forest
- Woody savanna
- Savanna
- Closed shrubland
- Open shrubland
- Grassland
- Cropland
- Barren or sparsely vegetated
- Urban or built-up
- Snow and ice
- Cropland/natural vegetation mosaic
- Wetland
- O Urban area with more than 3 million inhabitants

Yap Is.

FEDERATED STATES OF MICRONESIA

PALAU

Caroline Islands

Chuuk (Truk Is.)

Bismarck Archipelago

PAPUA NEW GUINEA

AUSTRALIA

O Sydn

O Melbourne

TASMANIA

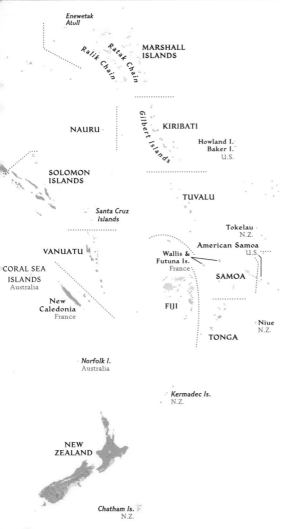

Enewetak
Atoll

Ratak Chain

Ralik Chain

**MARSHALL
ISLANDS**

NAURU

Gilbert Islands

KIRIBATI

Howland I.
Baker I.
U.S.

**SOLOMON
ISLANDS**

Santa Cruz
Islands

TUVALU

Tokelau
N.Z.

American Samoa
U.S.

VANUATU

Wallis &
Futuna Is.
France

SAMOA

CORAL SEA
ISLANDS
Australia

New
Caledonia
France

FIJI

Niue
N.Z.

TONGA

Norfolk I.
Australia

Kermadec Is.
N.Z.

**NEW
ZEALAND**

Chatham Is.
N.Z.

FEDERATED STATES
OF MICRONESIA

PALAU ●

MARSHALL
ISLANDS ●

NAURU ●

KIRIBATI ●

Howland I. ●
Baker I. ●
U.S.

PAPUA
NEW GUINEA
Af
H
Aw

SOLOMON
ISLANDS

TUVALU ●

Tokelau
N.Z.

VANUATU

American Samoa
Wallis & ●
Futuna Is. Fr.
Am

SAMOA

U.S.

Aw
Am
Bsh
Cwa
Aw

New Caledonia
France
Aw

FIJI

Norfolk Island
Australia

TONGA ●
Niue
N.Z.

AUSTRALIA

Aw
BSh
BSh
BWh
BSh
Csa
Csb

BSh

Cfa
Csa
Csb
Cfb

Cfa

NEW
ZEALAND
Cfb

TASMANIA

H

Climate Zones

(based on modified Köppen system)

Humid equatorial climate (A)
- No dry season (Af)
- Short dry season (Am)
- Dry winter (Aw)

Dry climate (B)
- Semiarid (BS) } h = hot
- Arid (BW) } k = cold

Humid temperate climate (C)
- No dry season (Cf)   a = hot summer
- Dry summer (Cs)   b = cool summer

Highland climate (H)
- Unclassified highlands

Water Availability

(millimeters per person
per year)

- More than 750
- 251–750
- 26–250
- Less than 26
- No data

POLAR REGIONS

ARCTIC DATA

AREA OF ARCTIC OCEAN: 14,700,000 sq km
(5,600,000 sq mi)

PERCENT OF EARTH'S WATER AREA:
4.1%

DEEPEST POINT: Molloy Deep
-5,669 m (-18,599 ft)

AVERAGE WINTER SEA ICE EXTENT:
15,500,000 sq km (6,000,000 sq mi)

AVERAGE SUMMER SEA ICE EXTENT:
6,500,000 sq km (2,500,000 sq mi)

ANTARCTICA DATA

AREA: 13,209,000 sq km (5,100,000 sq mi)

HIGHEST POINT: Vinson Massif
4,897 m (16,066 ft)

LOWEST POINT: Byrd Glacier
-2,870 m (-9,416 ft)

LOWEST RECORDED TEMPERATURE:
Vostok Research Station, Russia
-89.2°C (-128.6°F) July 21, 1983

HIGHEST RECORDED TEMPERATURE:
Vanda Station, New Zealand (currently
closed)   15°C (59°F) January 5, 1974

COLDEST PLACE ON EARTH:
Ridge A, annual average temperature
-74°C (-94°F)

Molloy Deep
-5,669 m (-18,599 ft)

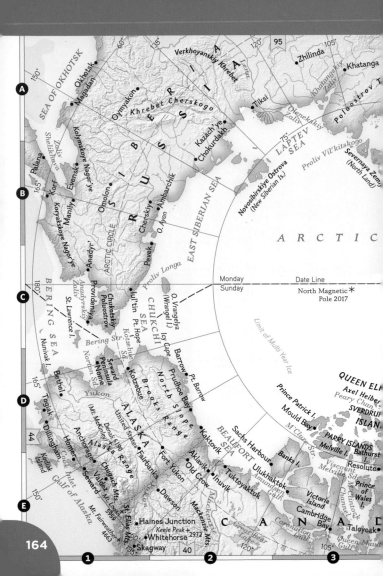

SEA OF OKHOTSK

Verkhoyanskiy Khrebet

Zhilinda

Khatanga

Okhotsk

Magadan

Oymyakon

Khrebet Cherskogo

Tiksi

Otenekskiy Zaliv

Khatangskiy Zaliv

Poluostrov T

Palana

Zaliv Shelikhova

Kolymskoye Nagor'ye

Kazách'ye

LAPTEV SEA

Proliv Vil'kitskogo

Severnaya Zem
(North Land)

Evensk

Omolon

Chokurdakh

Koryakskoye Nagor'ye

Manily

Chersky

Ambarchik

Novosibirskiye Ostrova
(New Siberian Is.)

S I B E R I A

R U S S I A

ARCTIC CIRCLE

Anadyr'

Omolon

O. Ayon

Pevek

EAST SIBERIAN SEA

A R C T I C

Proliv Longa

BERING SEA

Anadyrskiy Zaliv

Providemiya

Chukotskiy Poluostrov

St. Lawrence I.

O. Vrangelya
(Wrangel I.)

Monday
Sunday

Date Line

North Magnetic ✳
Pole 2017

Nunivak I.

Ivul'tin

Icy Cape

Pt. Hope

Kotzebue Sd.

CHUKCHI
SEA

Limit of Multi Year Ice

Bering Str.

Seward
Peninsula

Norton Sd.

Nome

Barrow

Pt. Barrow

Prudhoe Bay

North Slope

Prince Patrick I.

QUEEN EL

Axel Heiber

Peary Chan

Bethel

Kotzebue

Brooks Range

Mould Bay

SVERDRUP

ISLAN

Togiak

Yukon

Dillingham

44

ALASKA

Denali 6190
(Mt. McKinley)

United States

Fairbanks

Kaktovik

BEAUFORT
SEA

Sachs Harbour

Ulukhaktok

Banks I.

M'Clure Str.

Melville I.

PARRY ISLANDS

Bathurst

Resolute

Viscount

Melville

Sd.

Prince
of
Wales
I.

Kodiak
Island

Anchorage

Alaska Range

Chugach Mts.

Homer

Seward

Valdez

Mt. Logan
5959

Fort Yukon

Aklavik

Old Crow

Inuvik

Tuktoyaktuk

Amundsen
Gulf

Victoria
Island

Cambridge
Bay

Taloyoak

Gulf of Alaska

St. Elias Mts.

Mt. Fairweather 4663

Dawson

Mackenzie Mts.

Great Bear
Lake

Coronation
Gulf

Queen Maud
Gulf

Cook Inlet

Haines Junction

Keele Peak + 2972

Whitehorse

Skagway

40

C A N A D A

165°

150°

180°

60

135

120

95

105

75

150°

165°

150°

120°

105°

A

B

C

D

E

1

2

3

164

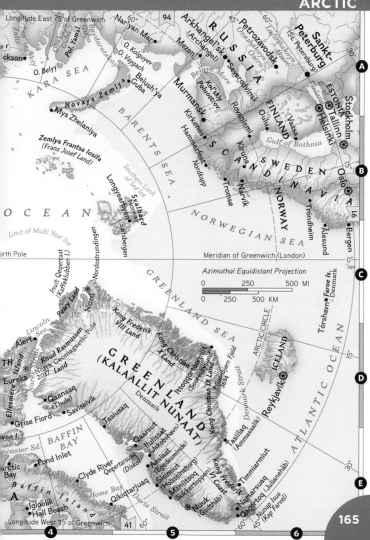

Longitude East 75 of Greenwich
Nar'yan Mar
94
Arkhangel'sk
(Archangel)
RUSSIA
Petrozavodsk
St. Petersburg
Sankt-
Peterburg
Belyy
O. Yamal
Mezen'
Onezhskoye
Ozero
Onega
Ladozhskoye
Ozero
A
ckson
O. Belyy
O. Kolguyev
Belush'ya
Guba
Severodvinsk
ESTONIA
Tallinn
KARA SEA
O. Vaygach
Murmansk
Kol'skiy
Poluostrov
Beloye More
Oulu
Vaasa
Stockholm
Helsinki
FINLAND
Novaya Zemlya
Mys Zhelaniya
Kirkenes
Rovaniemi
Kiruna
Gulf of
Bothnia
Oslo
B
BARENTS SEA
Hammerfest
Nordkapp
SWEDEN
NORWAY
16
15
Zemlya Frantsa Iosifa
(Franz Josef Land)
Tromso
Narvik
Trondheim
SCANDINAVIA
OCEAN
Southern Limit
of Sea Ice
Longyearbyen
Svalbard
Nordaust-
landet
Spitsbergen
NORWEGIAN
SEA
Ålesund
Bergen
Limit of Multi Year Ice
rth Pole
Nordostrundingen
Meridian of Greenwich (London)
0
C
Faroe Is.
Denmark
Tórshavn
Azimuthal Equidistant Projection
0   250   500 MI
0   250   500 KM
GREENLAND SEA
Inuit Qeqertaat
(Kaffeklubben I.)
Peary Land
Nord
Kong Frederik
VIII Land
ARCTIC CIRCLE
15
Lincoln
Sea
Alert
TH
Island
Knud Rasmussen
Land
North Geomagnetic Pole
2017
Eureka
Kane
Basin
Kong Christian
X Land
GREENLAND
(KALAALLIT NUNAAT)
Ittoqqortoormiit
(Scoresbysund)
Kong Frederik
Christian Field
Gunnbjørn Field
ICELAND
Reykjavik
ATLANTIC OCEAN
D
von I.
Qaanaaq
(Thule)
Denmark
Savissivik
Grise Fiord
Ellesmere
BAFFIN
BAY
Tasiusaq
Qaarsut
Ilulissat
(Jakobshavn)
Tasiilaq
(Ammassalik)
15
ncaster Sd.
Pond Inlet
Aasiaat
(Egedesminde)
Denmark Strait
Kangerlussuaq
rctic
Bay
Clyde River
Qeqertarsuaq
(Disko)
Sisimiut
(Holsteinsborg)
Kong Frederik
VI Coast
Timmiarmiut
30
E
A
Igloolik
Hall Beach
Qikiqtarjuaq
Home Bay
Maniitsoq
(Sukkertoppen)
Narsarsuaq
Qaqortoq (Julianehåb)
Baffin
Island
Baffin
Bay
Davis Strait
Nuuk
(Godthåb)
Nunap Isua
(Kap Farvel)
Longitude West 75 of Greenwich
41
60
45
4
5
6

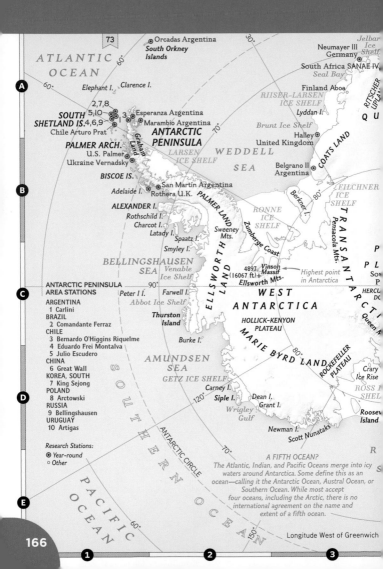

ATLANTIC
OCEAN

73

⊙ Orcadas Argentina
South Orkney Islands

Jelbar Ice Shelf

Neumayer III
Germany ⊙
South Africa SANAE IV

Seal Bay

60°

⊙A

Elephant I.   Clarence I.

Finland Aboa

RIISER-LARSEN
ICE SHELF

Lyddan I.

RITSCHER
UPLA

2,7,8
5,10 ⊙⊙⊙ 3
SOUTH
SHETLAND IS. 4,6,9 ⊙ 1

Esperanza Argentina
Marambio Argentina

Brunt Ice Shelf

Halley ⊙
United Kingdom

Q

Chile Arturo Prat

ANTARCTIC
PENINSULA

PALMER ARCH.
U.S. Palmer ⊙
Ukraine Vernadsky ⊙

LARSEN
ICE SHELF

WEDDELL

SEA

COATS LAND

Belgrano II ⊙
Argentina

Berkner I.

FILCHNER
ICE
SHELF

B

BISCOE IS.

Adelaide I. ⊙ San Martín Argentina
Rothera U.K.

PALMER LAND

RONNE
ICE
SHELF

Pensacola Mts.

ALEXANDER I.

Rothschild I.
Charcot I.
Latady I.

Smyley I.

Spaatz I.

BELLINGSHAUSEN
SEA

Venable
Ice Shelf

Sweeney
Mts.

Zumberge Coast

ELLSWORTH
LAND

4897
(16067 ft) ▲ Vinson
Massif

Ellsworth Mts.

Highest point
in Antarctica

TRANSANTARCTI

P
PL
So

HERCU

Queen M

ANTARCTIC PENINSULA
AREA STATIONS

ARGENTINA
1 Carlini
BRAZIL
2 Comandante Ferraz
CHILE
3 Bernardo O'Higgins Riquelme
4 Eduardo Frei Montalva
5 Julio Escudero
CHINA
6 Great Wall
KOREA, SOUTH
7 King Sejong
POLAND
8 Arctowski
RUSSIA
9 Bellingshausen
URUGUAY
10 Artigas

Research Stations:
⊙ Year-round
○ Other

90°

Peter I I.

Abbot Ice Shelf

Farwell I.

WEST
ANTARCTICA

HOLLICK-KENYON
PLATEAU

MARIE BYRD LAND

ROCKEFELLER
PLATEAU

Crary
Ice Rise

ROSS
SHEL

Thurston
Island

Burke I.

AMUNDSEN
SEA

GETZ ICE SHELF

Carney I.

Siple I.

Dean I.

Grant I.

Wrigley
Gulf

Newman I.
Scott Nunataks

Roosev
Island

R

SOUTHERN

ANTARCTIC CIRCLE

PACIFIC

OCEAN

A FIFTH OCEAN?
The Atlantic, Indian, and Pacific Oceans merge into icy
waters around Antarctica. Some define this as an
ocean—calling it the Antarctic Ocean, Austral Ocean, or
Southern Ocean. While most accept
four oceans, including the Arctic, there is no
international agreement on the name and
extent of a fifth ocean.

Longitude West of Greenwich

1   2   3

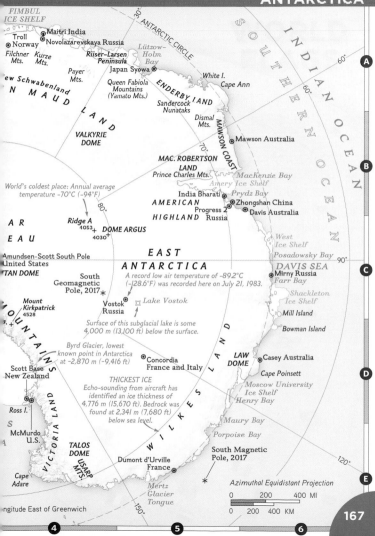

FIMBUL
ICE SHELF

30° ANTARCTIC CIRCLE

SOUTHERN OCEAN

INDIAN OCEAN

Troll
●Maitri India
◉Norway Novolazarevskaya Russia

Filchner
Mts. Kurze
Mts.

Riiser-Larsen
Peninsula

60°

A

New Schwabenland

Payer
Mts.

Japan Syowa ◉

White I.
Cape Ann

N M A U D L A N D

Queen Fabiola
Mountains
(Yamato Mts.)

ENDERBY
LAND

VALKYRIE
DOME

Sandercock
Nunataks

Dismal
Mts.

MAWSON COAST

◉Mawson Australia

60°

World's coldest place: Annual average
temperature -70°C (-94°F)

80°

MAC. ROBERTSON
LAND
Prince Charles Mts.

70°

MacKenzie Bay

B

Amery Ice Shelf

A R
E A U

Ridge A
4053+
◉DOME ARGUS
4030+

AMERICAN
HIGHLAND

India Bharati ◉
Progress 2
Russia

Prydz Bay
◉Zhongshan China
◉Davis Australia

West
Ice Shelf

Posadowsky Bay

90°

Amundsen-Scott South Pole
United States

E A S T
A N T A R C T I C A

A record low air temperature of -89.2°C
(-128.6°F) was recorded here on July 21, 1983.

DAVIS SEA

C

TAN DOME

South
Geomagnetic
Pole, 2017

◉Mirny Russia Farr Bay

Mount
Kirkpatrick
4528

Vostok
Russia ◉

☐Lake Vostok

Shackleton
Ice Shelf

Mill Island

M O U N T A I N S

Surface of this subglacial lake is some
4,000 m (13,100 ft) below the surface.

Bowman Island

Byrd Glacier, lowest
known point in Antarctica
at -2,870 m (-9,416 ft)

◉Concordia
France and Italy

LAW
DOME

◉Casey Australia

D

Scott Base
New Zealand

THICKEST ICE
Echo-sounding from aircraft has
identified an ice thickness of
4,776 m (15,670 ft). Bedrock was
found at 2,341 m (7,680 ft)
below sea level.

Cape Poinsett

Moscow University
Ice Shelf
Henry Bay

Ross I.

VICTORIA LAND

WILKES LAND

McMurdo
U.S.

Maury Bay

Porpoise Bay

TALOS
DOME

USARP
MTS.

Dumont d'Urville
France

South Magnetic
Pole, 2017

120°

Cape
Adare

Azimuthal Equidistant Projection

0   200   400 MI

0   200   400 KM

E

*

Mertz
Glacier
Tongue

150°

ngitude East of Greenwich

Afghanistan

 Albania

 Algeria

 Andorra

 Angola

 Antigua and Barbuda

 Argentina

 Armenia

 Australia

 Austria

 Azerbaijan

 Bahamas

 Bahrain

 Bangladesh

 Barbados

 Belarus

 Belgium

 Belize

 Benin

 Bhutan

 Bolivia

 Bosnia and Herzegovina

 Botswana

 Brazil

 Brunei

 Bulgaria

 Burkina Faso

 Burundi

 Cabo Verde

 Cambodia

 Cameroon

 Canada

 Central African Republic

 Chad

 Chile

China

Colombia

Comoros

Congo

Congo, Democralic
Republic of the

Costa Rica

Côte d'Ivoire
(Ivory Coast)

Croatia

Cuba

Cyprus

Czechia
(Czech Republic)

Denmark

Djibouti

Dominica

Dominican
Republic

Ecuador

Egypt

El Salvador

Equatorial Guinea

Eritrea

Estonia

Ethiopia

Fiji

Finland

France

Gabon

Gambia

Georgia

Germany

Ghana

Greece

Grenada

Guatemala

Guinea

Guinea-Bissau

Guyana

Haiti

Honduras

Hungary

Iceland

India

Indonesia

Iran

Iraq

Ireland

Israel

Italy

Jamaica

Japan

Jordan

Kazakhstan

Kenya

Kiribati

Korea, North

Korea, South

Kosovo

Kuwait

Kyrgyzstan

Laos

Latvia

Lebanon

Lesotho

Liberia

Libya

Liechtenstein

Lithuania

Luxembourg

Macedonia

Madagascar

Malawi

Malaysia

Maldives

Mali

Malta

Marshall Islands

Mauritania

Mauritius

Mexico

Micronesia

Moldova

Monaco

Mongolia

Montenegro

Morocco

Mozambique

Myanmar (Burma)

Namibia

Nauru

Nepal

Netherlands

New Zealand

Nicaragua

Niger

Nigeria

Norway

Oman

Pakistan

Palau

Panama

Papua
New Guinea

Paraguay

Peru

Philippines

Poland

Portugal

Qatar

Romania

Russia

Rwanda

Saint Kitts
and Nevis

Saint Lucia

Saint Vincent and
the Grenadines

Samoa

San Marino

Sao Tome
and Principe

Saudi Arabia

Senegal

Serbia

Seychelles

Sierra Leone

Singapore

Slovakia

Slovenia

Solomon Islands

Somalia

South Africa

South Sudan

Spain

Sri Lanka

Sudan

174

Suriname

Swaziland

Sweden

Switzerland

Syria

Tajikistan

Tanzania

Thailand

Timor-Leste
(East Timor)

Togo

Tonga

Trinidad and
Tobago

Tunisia

Turkey

Turkmenistan

Tuvalu

Uganda

Ukraine

United Arab
Emirates

United Kingdom

United States

Uruguay

Uzbekistan

Vanuatu

Vatican City

Venezuela

Vietnam

Yemen

Zambia

Zimbabwe

Adrar _____ mountain-s, plateau
Archipiélago _____ archipelago
Arquipélago _____ archipelago
Aylagy _____ gulf

Bāb _____ gate, strait
Bahía _____ bay
Bahr, Baḥr _____ bay, lake, river, sea
Baía, Baie _____ bay
Boğazı _____ strait
Boca _____ channel, mouth, river
Bögeni _____ reservoir
Bucht _____ bay
Burnu _____ cape, point

Cabo _____ cape
Canal _____ canal, channel, strait
Cap, Capo _____ cape
Catarata-s _____ cataract-s, waterfall-s
Cay-s, Cayo-s _____ island-s, key-s, shoal-s
Cerro-s _____ hill-s, peak-s
Cordillera _____ mountain chain

Dağı _____ mountain
Dağları _____ mountains
Dao _____ island
Daryācheh _____ lake, marshy lake
Dasht _____ desert, plain
Deniz, -i _____ sea
Desierto _____ desert
Do _____ island-s, rock-s
Doi _____ hill, mountain

Embalse _____ lake, reservoir
Emi _____ mountain, rock
Ensenada _____ bay, cove
Erg _____ sand dune region
Estrecho _____ strait

Feng _____ mount, peak
Fjeld _____ mountain, nunatak
Fjord-en _____ inlet, fjord

Gaoyuan _____ plateau
Gebel _____ mountain-s, range
Gezâir _____ islands
Gezîra-t, Gezîret _____ island, peninsula
Ghubb-at _____ bay, gulf
Gobi _____ desert
Gölü _____ lake

Golf-e, -o _____ gulf
Gora _____ mountain,-s
Got _____ point
Guba _____ bay, gulf

Hai _____ lake, sea

Haixia _____ channel, strait
Hamada, Ḥammādah _____ rocky desert
Hāyk' _____ lake, reservoir
He _____ canal, lake, river
Hu _____ lake, reservoir

Île-s, Ilha-s, Illa-s _____ island-s, islet-s
Isla-s _____ island-s
Isol-a, -e _____ island, isle
Istmo _____ isthmus

Jabal, Jebel _____ mountain-s, range
Jazā'ir, Jazīrat _____ island-s
Jezero _____ lake
Jiang _____ river, stream
Jibāl _____ hill, mountain, ridge
Jima _____ island-s, rock-s

Kap, Kapp _____ cape
Kaikyō _____ channel, strait
Kangri _____ mountain, peak
Kavīr _____ salt desert
Kepulauan _____ archipelago, islands
Khalîg, Khalīj _____ bay, gulf
Khrebet _____ mountain range
Ko _____ island, lake
Köli _____ lake
Kong _____ king, mountain
Körfezi _____ bay, gulf
Kūh, Kūhhā _____ mountain-s, range

Lac, Lac-ul _____ lake
Lae _____ cape, point
Lago, -a _____ lagoon, lake
Laguna-s _____ lagoon-s, lake-s
Laht _____ bay, gulf, harbor
Liedao _____ archipelago, islands
Ling _____ mountain-s, range

Man _____ bay
Mar, Mare _____ large lake, sea
Massif _____ mountain-s
Mesa, Meseta _____ plateau, tableland
Misaki _____ cape, peninsula, point
Mont-e, -s _____ mount, -ain, -s
Montagne-s _____ mount, -ain, -s

More _____ sea
Morro _____ bluff, headland, hill
Munkhafad _____ depression
Mys _____ cape

Nada _____ gulf, sea
Nafūd _____ area of dunes, desert
Nagor'ye _____ mountain range, plateau
Nevado-s _____ snow-capped mountain-s
Nord-re _____ north-ern
Nosy _____ island, reef, rock
Nunatak-s _____ peak-s surrounded by ice cap
Nur _____ lake, salt lake
Nuruu _____ mountain range, ridge
Nuur _____ lake

Ostrov,-a _____ island, -s
Ozer-o _____ lake, -s

Pampa-s _____ grassy plain-s
Pantanal _____ marsh, swamp
Pendi _____ basin
Península, Péninsule _____ peninsula
Phu _____ mountain
Pic, Pik _____ peak
Pico-s _____ peak-s
Playa _____ beach, inlet, shore
Planalto, Plato _____ plateau
Pointe _____ point
Poluostrov _____ peninsula
Ponta _____ cape, point
Proliv _____ strait
Puerto _____ bay, pass, port
Pulau _____ island
Punta _____ point

Qum _____ desert, sand
Qundao _____ archipelago, islands

Raas _____ cape, point
Ras, Râs, Ra's _____ cape
Represa _____ reservoir
Rettō _____ chain of islands
Rio, Río _____ river
Rocas _____ rocks

Sahara, Şaḥrā' _____ desert
Salar _____ salt flat
Salina _____ salt pan
Salin-as, -es _____ salt flat-s, salt marsh-es
Sankt _____ saint

Sanmaek _____ mountain range
Sa~o _____ saint
Sarīr _____ gravel desert
Selat _____ strait
Semenanjung _____ peninsula
Serra _____ range of hills or mountains
Shamo _____ desert
Shan _____ island-s, mountain-s, range
Shatt, Shaṭṭ _____ large river
Shima _____ island-s, rock-s
Shotō _____ archipelago
Shott _____ intermittent salt lake
Shyghanaghy _____ bay, gulf
Sierra-s _____ mountain range-s
Stretto _____ strait
Sud _____ south
Suidō _____ channel, strait
Sund _____ sound, strait

Tanjung _____ cape, point
Tassili _____ plateau, upland
Taüy _____ hills, mountains
Teluk _____ bay
Tierra _____ land, region
Tō _____ islands, rocks

Uul _____ mountain, range

Vodokhranilishche _____ reservoir
Volcan, Volcán _____ volcano
Vostochn-yy, -aya, -oye _____ eastern

Wadī _____ valley, watercourse
Wan _____ bay, gulf

Xia _____ gorge, strait
Xiao _____ lesser, little

Yoma _____ mountain range
Yuzhn-yy, -aya, -oye _____ southern

Zaki _____ cape, point
Zaliv _____ bay, gulf
Zapadn-yy, -aya, -oye _____ western
Zatoka _____ bay, gulf
Zemlya _____ land
Zhotasy _____ mountains

PLACE-NAME INDEX

Using the Index

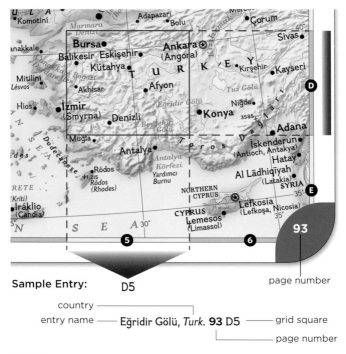

Sample Entry: D5

page number

country
entry name —— Eğridir Gölü, *Turk.* **93** D5 —— grid square
page number

The following system is used to locate a place on a map in this atlas. The bold-face type after an entry refers to the page on which the place-name is found. The letter-number combination refers to the grid on which the place-name is located. The edge of each map is marked with numbers and letters, dividing the map into a grid of squares. A name may appear on several maps, but the index lists only the best presentation. Some entries include a description, as in "Apure, *river, Venez.* **68** B3." In languages other than English, the description of a physical feature may be part of the name: For instance, "Shotō" in "Izu Shotō, *Japan* **121** D6" means "islands." When a feature or place can be referred to by more than one name, both may appear in the index with cross-references. For example, the entry for Cairo, Egypt reads "Cairo *see* El Qâhira, *Egypt* **140** A2." That entry is: "El Qâhira (Cairo), *Egypt* **140** A2."

B

Barlee, Lake, *W. Austral., Austral.* **150** D2

Barnaul, *Russ.* **118** A2

Barquisimeto, *Venez.* **68** A3

Barra, *Braz.* **71** C5

Barra, Ponta da, *Mozambique* **145** C4

Barra do São Manuel, *Braz.* **69** E5

Barranquilla, *Col.* **68** A2

Barra Point, *Mozambique* **129** E4

Barreiras, *Braz.* **71** C5

Barreirinhas, *Braz.* **71** A5

Barrow, *Alas., U.S.* **44** A3

Barrow, Point, *Alas., U.S.* **44** A3

Barrow Creek, *N. Terr., Austral.* **149** B4

Barrow Island, *W. Austral., Austral.* **150** B1

Bartolomeu Dias, *Mozambique* **145** C4

Baruun Urt, *Mongolia* **119** B4

Barysaw, *Belarus* **91** E5

Basankusu, *Dem. Rep. of the Congo* **142** A3

Bascuñán, Cabo, *Chile* **72** A1

Bashi Channel, *Philippines, Taiwan* **119** E5

Basilan, *island, Philippines* **123** C4

Basoko, *Dem. Rep. of the Congo* **142** A3

Basra *see* Al Başrah, *Iraq* **111** D4

Bassas da India, *islands, Fr.* **145** C5

Basseterre, *St. Kitts & Nevis* **52** B3

Basse-Terre, *island, Guadeloupe, Fr.* **52** B4

Basse-Terre, *Guadeloupe, Fr.* **52** B4

Bass Strait, *Austral.* **150** E3

Bastia, *Fr.* **89** E5

Basuo *see* Dongfang, *China* **119** E4

Bata, *Eq. Guinea* **142** A1

Batabanó, Golfo de, *Cuba* **50** C1

Batangas, *Philippines* **123** B4

Batan Islands, *Philippines* **123** A4

Batanme, *island, Indonesia* **123** D5

Bathurst, *N.B., Can.* **41** D5

Bathurst, *N.S.W., Austral.* **149** E5

Bathurst Inlet, *Nunavut, Can.* **40** B3

Bathurst Island, *N. Terr., Austral.* **150** A3

Bathurst Island, *Nunavut, Can.* **40** A3

Batna, *Alg.* **139** A4

Baton Rouge, *La., U.S.* **43** D4

Batu, Kepulauan, *Indonesia* **122** D1

Batumi, *Rep. of Ga.* **110** A3

Baubau, *Indonesia* **123** E4

Bauchi Plateau, *Nigeria* **128** C2

Bauld, Cape, *Nfld. & Lab., Can.* **41** C6

Bauru, *Braz.* **71** D4

Bawku, *Ghana* **138** D3

Bayamo, *Cuba* **50** D3

Bayan Har Shan, *China* **118** C3

Bayanhongor, *Mongolia* **118** B3

Baydaratskaya Guba, *Russ.* **165** A4

Baydhabo (Baidoa), *Somalia* **141** E4

Bay Islands, *Hond.* **31** E3

Baykal, Ozero, *Russ.* **119** A4

Baykonur *see* Bayqongyr, *Kaz.* **114** C3

Baykonur Cosmodrome, *Kaz.* **114** C3

Bayqongyr (Baykonur, Leninsk), *Kaz.* **114** C3

Beata, Cabo, *Dom. Rep.* **51** E5

Beaufort Sea, *Arctic Oc.* **164** D2

Beaufort West, *S. Af.* **144** E2

Beaumont, *Tex., U.S.* **49** A4

Béchar, *Alg.* **138** A3

Beddouza, Cap, *Mor.* **138** A2

Beddouza, Cape, *Mor.* **128** B1

Behbahān, *Iran* **111** D5

Behchokò, *N.W.T., Can.* **40** B2

Beida *see* Al Bayḑā', *Lib.* **139** A5

Beihai, *China* **119** E4

Beijing, *China* **119** B5

Beira, *Mozambique* **145** C4

Beirut *see* Beyrouth, *Lebanon* **110** C3

Bejaïa (Bougie), *Alg.* **89** F5

Bekasi, *Indonesia* **122** E2

Bela-Bela, *S. Af.* **144** C3

Belagavi (Belgaum), *India* **116** D3

Belarus, *Eur.* **77** C4

Belaya Gora, *Russ.* **95** B5

Belcher Islands, *Nunavut, Can.* **41** C4

Beledweyne, *Somalia* **141** E4

Belém, *Braz.* **71** A5

Belén, *Arg.* **72** A2

Belfast, *U.K.* **88** B2

Belgaum *see* Belagavi, *India* **116** D3

Belgium, *Eur.* **76** C2

Belgrade *see* Beograd, *Serb.* **92** B3

Belgrano II, *research station, Antarctica* **166** B3

Belitung, *island, Indonesia* **122** D2

Belize, *N. Amer.* **29** E3

Belize City, *Belize* **49** D5

Bellary *see* Ballari, *India* **116** D3

Bir-Bos

Bosnia and Herzegovina, *Eur.* **76** D3

Bosobolo, *Dem. Rep. of the Congo* **142** A3

Bosporus *see* İstanbul Boğazı, *strait, Turk.* **93** C5

Bossangoa, *Cen. Af. Rep.* **139** E5

Boston, *Mass., U.S.* **43** B6

Botany Bay, *N.S.W., Austral.* **151** E6

Bothnia, Gulf of, *Fin., Sw.* **90** C3

Botswana, *Af.* **127** E3

Bouaké, *Côte d'Ivoire* **138** E2

Bouar, *Cen. Af. Rep.* **139** E5

Bougainville, *island, P.N.G.* **160** C2

Bougainville, Cape, *W. Austral., Austral.* **150** A3

Bougie *see* Bejaia, *Alg.* **89** F5

Bouira, *Alg.* **89** F4

Boujdour, *W. Sahara, Mor.* **138** B1

Boujdour, Cape, *W. Sahara, Mor.* **128** B1

Boulder, *Colo., U.S.* **42** C3

Bourke, *N.S.W., Austral.* **149** D5

Bowen, *Qnsld., Austral.* **149** B5

Bowman Island, *Antarctica* **167** D6

Boyarka, *Russ.* **95** C4

Bozashchy Tübegi, *Kaz.* **114** C1

Bozeman, *Mont., U.S.* **42** B2

Brades, *Montserrat, U.K.* **52** B3

Braga, *Port.* **89** E2

Brahmapur, *India* **117** C4

Brahmaputra *see* Yarlung Zangbo, *river, Bangladesh, China, India* **117** B5

Brăila, *Rom.* **93** B4

Branco, *river, Braz.* **69** C4

Brandberg, *peak, Namibia* **144** C1

Brandon, *Man., Can.* **40** D3

Brandvlei, *S. Af.* **144** D2

Brasília, *Braz.* **71** C5

Brasileiro, Planalto, *Braz.* **71** D5

Brașov, *Rom.* **93** B4

Bratislava (Pressburg), *Slovakia* **92** A3

Bratsk, *Russ.* **99** B4

Braunschweig (Brunswick), *Ger.* **88** B5

Bravo del Norte, Río *see* Grande, Rio, *Mex.* **48** A2

Brazil, *S. Amer.* **56** C3

Brazilian Highlands, *Braz.* **58** C4

Brazzaville, *Congo* **142** B2

Brecknock, Península, *Chile* **73** F2

Bredy, *Russ.* **114** A3

Brejo, *Braz.* **71** A5

Bremen, *Ger.* **91** E1

Brenner Pass, *Aust., It.* **92** A1

Brest, *Belarus* **91** F4

Brest, *Fr.* **88** C3

Bria, *Cen. Af. Rep.* **139** E6

Bridgetown, *Barbados* **53** D5

Brighton, *U.K.* **88** C3

Brindisi, *It.* **92** D3

Brisbane, *Qnsld., Austral.* **149** D6

Bristol, *U.K.* **88** B3

Bristol Bay, *Alas., U.S.* **44** C2

British Columbia, *province, Can.* **40** C1

British Isles, *Ire., U.K.* **78** B2

British Virgin Islands, *possession, U.K.* **52** A1

Britstown, *S. Af.* **144** D2

Brive, *Fr.* **89** D4

Brno, *Czech Rep.* **92** A2

Broken Bay, *N.S.W., Austral.* **149** E6

Broken Hill, *N.S.W., Austral.* **149** D4

Broken Hill *see* Kabwe, *Zambia* **143** E4

Brønnøysund, *Nor.* **90** B2

Brooks Range, *Alas., U.S.* **44** A3

Broome, *W. Austral., Austral.* **148** B2

Broughton Island *see* Qikiqtarjuaq, *Nunavut, Can.* **41** B5

Brownsville, *Tex., U.S.* **42** E3

Bruce, Mount, *W. Austral., Austral.* **150** C2

Brunei, *Asia* **99** E5

Brunswick *see* Braunschweig, *Ger.* **88** B5

Brunswick Bay, *W. Austral., Austral.* **150** A2

Brunt Ice Shelf, *Antarctica* **166** A3

Brussels *see* Bruxelles, *Belg.* **76** C2

Bruxelles (Brussels), *Belg.* **76** C2

Bryce Canyon National Park, *U.S.* **46** C2

Bucaramanga, *Col.* **68** B2

Buchanan, *Liberia* **138** E2

Bucharest *see* București, *Rom.* **93** B4

Buckland Tableland, *Qnsld., Austral.* **151** C5

București (Bucharest), *Rom.* **93** B4

Budapest, *Hung.* **92** A3

Buenaventura, *Col.* **68** B1

Buenaventura, Bahía de, *Col.* **56** A1

Buenaventura Bay, *Col.* **58** B1

Buenos Aires, *Arg.* **72** B4

Buenos Aires, Lago, *Arg.* **73** E2

Buffalo, *N.Y., U.S.* **43** B5

Bug, *river, Eur.* **91** F4

Bügür *see* Luntai, *China* **115** C6

Bujumbura, *Burundi*
143 B4

Bukama, *Dem. Rep. of the Congo* 143 D4

Bukavu, *Dem. Rep. of the Congo* 143 B4

Bukhara *see* Buxoro, *Uzb.*
114 D3

Bulawayo, *Zimb.* 144 C3

Bulgan, *Mongolia* 119 A4

Bulgaria, *Eur.* 77 D4

Bumba, *Dem. Rep. of the Congo* 142 A3

Bunbury, *W. Austral., Austral.* 148 E1

Bundaberg, *Qnsld., Austral.*
149 C6

Bungo Suidō, *Japan* 121 D4

Bunia, *Dem. Rep. of the Congo* 143 A4

Buon Me Thuot, *Vietnam*
122 B2

Bunia, *Dem. Rep. of the Congo* 143 A4

Buon Me Thuot, *Vietnam*
122 B2

Burao *see* Burco, *Somaliland* 141 D5

Buraydah, *Saudi Arabia*
112 B3

Burco (Burao), *Somaliland*
141 D5

Burgas, *Bulg.* 93 C4

Burgos, *Sp.* 89 E3

Burica, Punta, *Pan.* 49 E5

Burke Island, *Antarctica*
166 C2

Burketown, *Qnsld., Austral.*
149 B4

Burkina Faso, *Af.* 126 C2

Burma *see* Myanmar, *Asia*
99 D4

Burnie, *Tas., Austral.*
148 E3

Burnside, Lake, *W. Austral., Austral.* 150 C2

Burrendong Reservoir, *N.S.W., Austral.* 151 D5

Bursa, *Turk.* 110 A1

Bûr Safâga, *Egypt* 110 E2

Bûr Sa'îd, *Egypt* 140 A2

Buru, *island, Indonesia*
123 D4

Burultokay *see* Fuhai, *China*
115 B6

Burundi, *Af.* 127 D4

Busan (Pusan), *S. Kor.*
120 D3

Būshehr, *Iran* 111 D5

Businga, *Dem. Rep. of the Congo* 142 A3

Busselton, *W. Austral., Austral.* 148 E1

Buta, *Dem. Rep. of the Congo* 142 A3

Butembo, *Dem. Rep. of the Congo* 143 B4

Buton, *island, Indonesia*
123 E4

Butte, *Mont., U.S.* 42 B2

Butuan, *Philippines* 123 C4

Buxoro (Bukhara), *Uzb.*
114 D3

Buyant-Uhaa *see* Saynshand, *Mongolia*
119 B4

Bydgoszcz, *Pol.* 91 E3

Bylot Island, *Nunavut, Can.*
41 A4

Byron, Cape, *N.S.W., Austral.* 151 D6

C

Cabinda, *region, Angola*
142 C1

Cabinda, *Angola* 142 C1

Cabo San Lucas, *Mex.*
48 B1

Cáceres, *Braz.* 70 C3

Cáceres, *Sp.* 89 E2

Cachimbo, *Braz.* 69 E5

Cadale, *Somalia* 141 E5

Cádiz, *Sp.* 89 F2

Cagayan de Oro, *Philippines* 123 C4

Cagliari, *It.* 92 D1

Cahora Bassa, Lago de, *Mozambique* 143 E4

Caicos Islands, *Turks & Caicos Is., U.K.* 51 C4

Caicos Passage, *Bahamas, Turks & Caicos Is.* 51 C4

Cairns, *Qnsld., Austral.*
149 B5

Cairo *see* El Qâhira, *Egypt*
140 A2

Cajamarca, *Peru* 68 E1

Calais, *Fr.* 88 C4

Calama, *Braz.* 69 E4

Calama, *Chile* 70 D2

Calamar, *Col.* 68 C2

Calamian Group, *Philippines*
123 B4

Călăraşi, *Rom.* 93 B4

Calbayog, *Philippines*
123 B4

Calcanhar, Point, *Braz.*
58 B5

Calcanhar, Ponta do, *Braz.*
56 B5

Calçoene, *Braz.* 69 C6

Calcutta *see* Kolkata, *India*
117 C5

Calgary, *Alberta, Can.*
40 D2

Cali, *Col.* 68 C2

California, *state, U.S.* 42 C1

California, Golfo de, *Mex.*
48 A1

California, Gulf of, *Mex.*
31 D1

Callao, *Peru* 70 C1

Caloundra, *Qnsld., Austral.*
149 C6

Calulo, *Angola* 142 D2

Calvinia, *S. Af.* 144 E2

Camabatela, *Angola* 144 A1

Camacupa, *Angola* 144 A2

Camagüey, *Cuba* 50 C3

Camagüey, Archipiélago de, *Cuba* 50 C3

Camarones, *Arg.* 73 D3

Cambodia, *Asia* 99 D4

Cambridge Bay, *Nunavut, Can.* 40 B2

Cambundi-Catembo, *Angola*
144 A2

Camden Bay, *Alas., U.S.*
44 A4

Cameroon, *Af.* 126 C3

Cameroon Mountain, *Cameroon* 142 A1

Camiri, *Bol.* 70 D3

Camocim, *Braz.* 71 A6

Catalina, Punta, *Arg., Chile*
73 F2

Catamarca, *Arg.* **72** A2

Catanduanes, *island,
Philippines* **123** B4

Catania, *It.* **92** D2

Catanzaro, *It.* **92** D2

Catete, *Angola* **142** D2

Catingas, *region, Braz.*
58 B3

Cat Island, *Bahamas* **50** B3

Caubvick, Mount *see*
D'Ibervile, Mont, *Can.*
41 C5

Cauca, *river, Col.* **68** B2

Caucasus Mountains,
Azerb., Rep. of Ga., Russ.
110 A3

Caura, *river, Venez.* **69** B4

Cauto, *river, Cuba* **50** C3

Caxias, *Braz.* **71** B5

Caxias do Sul, *Braz.* **71** E4

Caxito, *Angola* **144** A1

Cayenne, *Fr. Guiana, Fr.*
69 B6

Cayman Brac, *island,
Cayman Is., U.K.* **50** D2

Cayman Islands, *possession,
U.K.* **50** D2

Cazombo, *Angola* **144** A2

Cebu, *island, Philippines*
123 B4

Cebu, *Philippines* **123** B4

Ceduna, *S. Austral., Austral.*
149 D4

Celebes *see* Sulawesi,
island, Indonesia **123** D4

Celebes Sea, *Indonesia,
Philippines* **123** C4

Celtic Sea, *Eur.* **78** C2

Cenderawasih, Teluk,
Indonesia **123** D6

Central, Cordillera, *Col.*
68 C2

Central, Cordillera, *Dom.
Rep.* **51** D5

Central, Massif, *Fr.* **78** D2

Central African Republic,
Af. **126** C3

Central America, *region, N.
Amer.* **31** E3

Central Lowland, *U.S.*
31 D3

Central Lowlands, *Austral.*
151 B4

Central Siberian Plateau,
Russ. **101** B4

Ceram *see* Seram, *island,
Indonesia* **123** D5

Ceram Sea, *Indonesia*
123 D5

Cerro de Pasco, *Peru*
70 C1

Ceuta, *Sp.* **89** F2

Chābahār, *Iran* **113** B6

Chacao, Canal de, *Chile*
73 D1

Chad, *Af.* **126** C3

Chad, Lake, *Af.* **139** D5

Chagda, *Russ.* **95** C5

Chagos Archipelago,
possession, U.K. **98** E3

Chaman, *Pak.* **116** A2

Chambi, Jebel ech, *Tun.*
139 A4

Chañaral, *Chile* **72** A1

Chandigarh, *India* **116** B3

Chandrapur, *India* **117** C4

Changchun, *China* **119** B5

Chang Jiang (Yangtze),
China **119** D4

Changjin Reservoir (Chosin
Reservoir), *N. Kor.*
120 B2

Changsan-got, *cape, N. Kor.*
120 C1

Changsha, *China* **119** D5

Changwon, *S. Kor.* **120** D2

Changyŏn, *N. Kor.* **120** C1

Channel Islands, *U.K.*
88 C3

Channel Country, *Austral.*
151 C4

Channel Islands, *Calif., U.S.*
42 C1

Channel Islands National
Park, *U.S.* **46** C1

Chany, Ozero, *Russ.* **115** A5

Chapayevsk, *Russ.* **114** A2

Chappal Waddi, *peak,
Cameroon, Nigeria*
139 E4

Charcot Island, *Antarctica*
166 B2

Chari, *river, Chad* **139** D5

Chärjew *see* Türkmenabat,
Turkm. **115** D3

Charles, Cape, *U.S.* **31** D4

Charles Point, *N. Terr.,
Austral.* **150** A3

Charleston, *S.C., U.S.*
43 D5

Charleston, *W. Va., U.S.*
43 C5

Charlestown, *St. Kitts &
Nevis* **52** B3

Charleville, *Qnsld., Austral.*
149 C5

Charlotte, *N.C., U.S.* **43** C5

Charlotte Amalie, *U.S.
Virgin Is., U.S.* **52** A1

Charlotte Harbor, *U.S.*
50 A1

Charlottetown, *P.E.I., Can.*
41 D6

Charlotteville, *Trin. &
Tobago* **53** F5

Charters Towers, *Qnsld.,
Austral.* **149** B5

Chartres, *Fr.* **88** C4

Chasŏng, *N. Kor.* **120** A2

Chatham Islands, *N.Z.*
161 E4

Chaves, *Braz.* **69** C6

Chech, 'Erg, *Alg., Mali*
138 B3

Cheduba Island, *Myanmar*
117 C5

Chegutu, *Zimb.* **144** B3

Cheekha Dar, *peak, Iran,
Iraq* **111** B4

Chengchow *see* Zhengzhou,
China **119** B5

Chengdu, *China* **119** D4

Chennai (Madras), *India*
117 D4

Cheonan, *S. Kor.* **120** D2

Cheongju, *S. Kor.* **120** D2

Chelm, *Pol.* **91** F4

Chelyabinsk, *Russ.* **94** D2

Chemnitz, *Ger.* **91** F2

Cheju *see* Jeju, *S. Kor.*
120 E2

Cherbourg-Octeville, *Fr.*
 88 C3
Cherlak, *Russ.* **115** A5
Chernihiv, *Ukr.* **91** F5
Chernivtsi, *Ukr.* **93** A4
Chernyakhovsk, *Russ.*
 91 E3
Cherskiy, *Russ.* **95** B5
Cherskogo, Khrebet, *Russ.*
 95 B5
Chesapeake Bay, *U.S.*
 43 C6
Chesha Bay, *Russ.* **79** A5
Cheshskaya Guba, *Russ.*
 77 A5
Chesterfield Inlet, *Can.*
 30 B3
Chesterfield Inlet, *Nunavut,*
 Can. **40** C3
Chetumal, *Mex.* **49** C5
Ch'ew Bahir, *lake, Eth.*
 143 A6
Cheyenne, *Wyo., U.S.*
 42 C3
Chiang Mai, *Thai.* **122** A1
Chibemba, *Angola* **144** B1
Chibougamau, *Que., Can.*
 41 D5
Chicago, *Ill., U.S.* **43** B4
Chichagof Island, *Alas., U.S.*
 44 C3
Chiclayo, *Peru* **68** E1
Chico, *river, Arg.* **73** D2
Chigubo, *Mozambique*
 145 C4
Chihuahua, *Mex.* **48** B2
Chile, *S. Amer.* **57** E2
Chillán, *Chile* **72** C1
Chiloé, Isla Grande de,
 Chile **73** D1
Chilpancingo, *Mex.* **48** D3
Chilwa, Lake, *Malawi,*
 Mozambique **143** E5
Chimborazo, *peak, Ecua.*
 58 B1
Chimbote, *Peru* **68** E1
Chimboy, *Uzb.* **114** D3
Chimoio, *Mozambique*
 145 B4
China, *Asia* **99** C4

Chinde, *Mozambique*
 145 B4
Chindwin, *river, Myanmar*
 117 C5
Chingola, *Zambia* **143** D4
Chinhoyi, *Zimb.* **144** B3
Chipata, *Zambia* **143** D5
Chirinda, *Russ.* **95** C4
Chisasibi, *Que., Can.* **41** D4
Chisimayu *see* Kismaayo,
 Somalia **141** F4
Chişinău, *Mold.* **93** A5
Chita, *Russ.* **95** D5
Chitado, *Angola* **144** B1
Chitato (Dundo), *Angola*
 142 C3
Chitré, *Pan.* **49** E6
Chittagong, *Bangladesh*
 117 C5
Chitungwiza, *Zimb.* **145** B4
Chivilcoy, *Arg.* **72** B4
Chlef, *Alg.* **89** F4
Chokurdakh, *Russ.* **164** B2
Choma, *Zambia* **143** B3
Ch'ŏngjin, *N. Kor.* **120** A3
Chongqing, *China* **119** D4
Chonos, Archipiélago de
 los, *Chile* **73** D1
Chonos Archipelago, *Chile*
 59 E2
Chornobyl', *Ukr.* **91** F5
Ch'osan, *N. Kor.* **120** B2
Chosin Reservoir *see*
 Changjin Reservoir, *N.*
 Kor. **120** B2
Choybalsan, *Mongolia*
 119 A4
Christchurch, *N.Z.* **160** E3
Christiansted, *U.S. Virgin*
 Is., U.S. **52** A1
Christmas Island, *Austral.*
 122 E2
Christmas Island *see*
 Kiritimati, *Kiribati* **161** B5
Chubut, *river, Arg.* **73** D2
Chudovo, *Russ.* **91** D5
Chugach Mountains, *Alas.,*
 U.S. **44** B3
Chukchi Peninsula, *Russ.*
 101 A5

Chukchi Sea, *Arctic Oc.*
 164 C1
Chukotskiy Poluostrov,
 Russ. **164** C1
Chul'man, *Russ.* **95** C5
Chumikan, *Russ.* **95** C5
Chuncheon, *S. Kor.* **120** C2
Chur, *Switz.* **92** A1
Churchill, *Man., Can.*
 40 C3
Churchill, Cape, *Can.*
 30 B3
Chuuk (Truk Islands),
 F.S.M. **160** B2
Ciego de Ávila, *Cuba*
 50 C2
Cienfuegos, *Cuba* **50** C2
Cincinnati, *Ohio, U.S.*
 43 C5
Cinto, Monte, *Fr.* **92** C1
Cistern Point, *Bahamas*
 50 B3
Ciudad Acuña, *Mex.* **48** A3
Ciudad Bolívar, *Venez.*
 69 B4
Ciudad del Carmen, *Mex.*
 49 C4
Ciudad Guayana, *Venez.*
 69 A4
Ciudad Juárez, *Mex.*
 48 A2
Ciudad Madero, *Mex.*
 48 C3
Ciudad Obregón, *Mex.*
 48 B2
Ciudad Valles, *Mex.* **48** C3
Ciudad Victoria, *Mex.*
 48 B3
Clarence Island, *Antarctica*
 166 A1
Clarines, *Venez.* **53** F1
Cleveland, *Ohio, U.S.*
 43 B5
Clipperton, *island, Pacific*
 Oc. **48** E1
Cloncurry, *river, Qnsld.,*
 Austral. **151** B4
Cloncurry, *Qnsld., Austral.*
 149 B5
Cloncurry Plateau, *Qnsld.,*
 Austral. **151** B4

Clu-Cor

Córdoba, Sierras de, *Arg.*
72 B3

Cordova, *Alas., U.S.* **44** B4

Corfu *see* Kérkira, *island, Gr.* **92** D3

Cork (Corcaigh), *Ire.*
88 B2

Corner Brook, *Nfld. & Lab., Can.* **41** D6

Coronado Bay, *Costa Rica*
31 F3

Coronation Gulf, *Nunavut, Can.* **40** B2

Coropuna, Nevado, *Peru*
70 CI

Corpus Christi, *Tex., U.S.*
42 E3

Corrientes, *Arg.* **72** A4

Corrientes, *river, Ecua., Peru* **68** D2

Corrientes, Cabo, *Col.*
68 BI

Corrientes, Cape, *Col.*
58 AI

Corse, Cap, *Fr.* **92** CI

Corsica, *island, Fr.* **92** CI

Çoruh, *Turk.* **110** A3

Çorum, *Turk.* **110** A2

Corumbá, *Bol.* **70** D3

Cosenza, *It.* **92** D2

Cosmoledo Group, *Seychelles* **145** A6

Costa Rica, *N. Amer.* **29** F3

Côte d'Ivoire (Ivory Coast), *Af.* **126** CI

Cotonou, *Benin* **138** E3

Cottica, *Suriname* **69** B6

Cowan, Lake, *W. Austral., Austral.* **150** D2

Coxim, *Braz.* **71** D4

Cozumel, Isla, *Mex.* **49** C5

Cradock, *S. Af.* **144** E3

Craiova, *Rom.* **93** B4

Crary Ice Rise, *Antarctica*
166 D3

Crater Lake National Park, *U.S.* **46** BI

Crateús, *Braz.* **71** B6

Crete (Kríti), *island, Gr.*
93 E4

Crete, Sea of, *Gr.* **93** E4

Creus, Cap de, *Sp.* **89** E4

Crimea, *peninsula, Ukr.*
79 D5

Croatia, *Eur.* **76** D3

Crooked Island, *Bahamas*
51 C4

Crooked Island Passage, *Bahamas* **51** C4

Croydon, *Qnsld., Austral.*
149 B5

Cruz, Cabo, *Cuba* **50** D3

Cruzeiro do Sul, *Braz.*
68 E2

Crystal Mountains, *Af.*
142 BI

Cuamba, *Mozambique*
145 B4

Cuando, *river, Angola, Namibia, Zambia* **142** E3

Cuango, *river, Angola, Dem. Rep. of the Congo* **144** AI

Cuanza, *river, Angola*
142 D2

Cuba, *N. Amer.* **29** E4

Cúcuta, *Col.* **68** B2

Cuenca, *Ecua.* **68** DI

Cuenca, *Sp.* **89** E3

Cuiabá, *Braz.* **71** C4

Cuito, *river, Angola* **142** E3

Cuito Cuanavale, *Angola*
142 E2

Culebra, *island, P.R., U.S.*
52 AI

Culiacán, *Mex.* **48** B2

Culver, Point, *W. Austral., Austral.* **150** D3

Cumaná, *Venez.* **69** A4

Cumanacoa, *Venez.* **53** F2

Cumberland Peninsula, *Nunavut, Can.* **41** B5

Cumberland Sound, *Nunavut, Can.* **41** B5

Cuneo, *It.* **89** D5

Cunnamulla, *Qnsld., Austral.*
149 D5

Cupica, Golfo de, *Col.*
68 BI

Curaçao, *possession, Neth.*
53 D2

Curaray, *river, Ecua., Peru*
68 D2

Curicó, *Chile* **72** BI

Curitiba, *Braz.* **71** E4

Curralinho, *Braz.* **69** D6

Curvelo, *Braz.* **71** D5

Cusco, *Peru* **70** C2

Cuttack, *India* **117** C4

Cuyahoga Valley National Park, *U.S.* **47** B5

Cuyuni, *river, Guyana, Venez.* **69** B4

Cyclades *see* Kikládes, *islands, Gr.* **93** E4

Cyprus, *Eur.* **77** E4

Cyrenaica, *region, Lib.*
138 B5

Czechia (Czech Republic), *Eur.* **76** C3

Czech Republic *see* Czechia, *Eur.* **76** C3

D

Daegu (Taegu), *S. Kor.*
120 D2

Daejeon (Taejŏn), *S. Kor.*
120 D2

Dagelet *see* Ulleungdo, *island, S. Kor.* **120** C3

Dagö *see* Hiiumaa, *island, Est.* **91** D3

Da Hinggan Ling, *mountains, China* **119** B5

Dahlak Archipelago, *Eritrea*
140 C4

Dairen *see* Dalian, *China*
119 C5

Daitō Shotō, *Japan* **160** AI

Dakar, *Senegal* **138** DI

Dalandzadgad, *Mongolia*
119 B4

Da Lat, *Vietnam* **122** B2

Dalgaranga Hill, *W. Austral., Austral.* **150** DI

Dali, *China* **118** E3

Dalian (Dairen), *China*
119 C5

Dallas, *Tex., U.S.* **42** D3

Dall Island, *Alas., U.S.*
44 C5

Dal'negorsk, *Russ.* **95** D6

Dal'nerechensk, *Russ.*
121 A5

Daloa, *Côte d'Ivoire* 138 E2

Daly, *river, N. Terr., Austral.*
150 A3

Damar, *island, Indonesia*
123 E5

Damascus *see* Dimashq, *Syr.*
110 C2

Damāvand, Kūh-e, *Iran*
111 B5

Dampier Archipelago, *W. Austral., Austral.* 150 B1

Dampier Land, *W. Austral., Austral.* 150 B2

Danakil, *region, Eth.*
128 C5

Da Nang, *Vietnam* 122 B2

Dandong, *China* 119 B5

Danube, *river, Eur.* 79 D4

Dao Phu Quoc, *island, Vietnam* 122 B2

Da Qaidam, *China* 118 C3

Daqing, *China* 119 A5

Dardanelles *see* Çanakkale Boğazı, *strait, Turk.*
93 D4

Dar es Salaam, *Tanzania*
143 C6

Darfur, *region, Sudan*
141 D1

Darhan, *Mongolia* 119 A4

Darling, *river, N.S.W., Austral.* 151 D5

Darling Downs, *Qnsld., Austral.* 151 D5

Darling Range, *W. Austral., Austral.* 150 D2

Darnah (Derna), *Lib.*
139 A6

Dar Rounga, *hills, Cen. Af. Rep., Chad* 139 E6

Dartang *see* Baqên, *China*
118 D3

Darvel Bay, *Malaysia*
123 C4

Darwin, *N. Terr., Austral.*
148 A3

Daşoguz, *Turkm.* 114 D3

Datong, *China* 119 B4

Daugavpils, *Latv.* 91 D4

Dauphin, *Man., Can.*
40 D3

Davangere, *India* 116 D3

Davao, *Philippines* 123 C4

Davao Gulf, *Philippines*
123 C4

Davenport Range, *N. Terr., Austral.* 151 B4

David, *Pan.* 49 E5

Davis, *research station, Antarctica* 167 C6

Davis Sea, *Antarctica*
167 C6

Davis Strait, *Can., Greenland* 30 B4

Dawa, *river, Eth.* 143 A6

Dawei, *Myanmar* 122 B1

Dawmat al Jandal (Al Jawf), *Saudi Arabia*
112 A2

Dawson, *river, Qnsld., Austral.* 151 C6

Dawson, *Yukon, Can.*
40 A1

Dawson Creek, *B.C., Can.*
40 C2

Dayr az Zawr, *Syr.* 110 C3

Deadmans Cay, *Bahamas*
50 C3

Dead Sea, *Israel, Jordan*
110 D2

Dean Island, *Antarctica*
166 D2

Death Valley, *Calif., U.S.*
42 C1

Death Valley National Park, *U.S.* 46 C1

Débo, Lake, *Mali* 128 C2

Debrecen, *Hung.* 92 A3

Debre Mark'os, *Eth.*
141 D3

Deccan, *region, India*
116 D3

Deccan Plateau, *India*
100 D3

DeGrey, *river, W. Austral., Austral.* 150 B2

Deh Bīd, *Iran* 111 D5

Dejen, Ras, *Eth.* 141 D3

Delaware, *state, U.S.*
43 C6

Delgado, Cabo, *Mozambique*
145 A5

Delgado, Cape, *Mozambique, Tanzania*
129 D5

Delhi, *India* 116 B3

Delicias, *Mex.* 48 B2

Déline, *N.W.T., Can.* 40 B2

Dellys, *Alg.* 89 F4

De Long Mountains, *Alas., U.S.* 44 A2

Democratic Republic of the Congo, *Af.* 127 D3

Denakil, *region, Eth.*
141 D4

Denali (McKinley, Mount), *Alas., U.S.* 44 A4

Denali National Park and Preserve, *U.S.* 46 E2

Den Haag (The Hague), *Neth.* 88 B3

Denham, *W. Austral., Austral.* 148 C1

Denizli, *Turk.* 110 B1

Denmark, *Eur.* 76 C3

Denmark Strait, *Greenland, Ice.* 165 D6

Denpasar, *Indonesia*
122 E3

D'Entrecasteaux, Point, *W. Austral., Austral.* 150 E1

Denver, *Colo., U.S.* 42 C3

Deputatskiy, *Russ.* 95 B5

Derbent, *Russ.* 114 D1

Derby, *W. Austral., Austral.*
148 B2

Derna *see* Darnah, *Lib.*
139 A6

Dernberg, Cape, *Namibia*
144 D1

Derudeb, *Sudan* 140 C3

Derzhavīnsk, *Kaz.* 115 B4

Desē, *Eth.* 141 D4

Deseado, *river, Arg.* 73 E2

Desengaño, Punta, *Arg.*
73 E2

Des Moines, *Iowa, U.S.*
43 B4

Detroit, *Mich., U.S.* 43 B5

Devon Island, *Nunavut, Can.*
41 A4

Devonport, *Tas., Austral.*
148 E3

Deyang, *China* **119** D4

Dezfūl, *Iran* **111** C5

Dhaka, *Bangladesh* **117** C5

Dhamār, *Yemen* **112** E3

Dhofar *see* Z̧ufār, *region, Oman* **113** D5

Diamante, *river, Arg.* **72** B2

Diamantina, *Braz.* **71** D5

Diamantina, *river, Qnsld., Austral.* **151** C5

Diamond Head, *Hawai'i, U.S.* **45** E3

D'Iberville, Mont (Mount Caubvick), *Can.* **41** C5

Dibrugarh, *India* **117** B5

Dickson, *Russ.* **94** B3

Didao, *China* **119** A6

Diego de Almagro, *Chile* **72** A2

Diffa, *Niger* **139** D5

Dijon, *Fr.* **88** C4

Dika, Mys, *Russ.* **95** B4

Dili, *Timor-Leste* **123** E4

Dillingham, *Alas., U.S.* **44** C2

Dilolo, *Dem. Rep. of the Congo* **142** D3

Dimapur, *India* **117** B5

Dimitrovgrad, *Russ.* **114** A2

Diourbel, *Senegal* **138** D1

Dirē Dawa, *Eth.* **141** D4

Dirico, *Angola* **142** E3

Dirk Hartog Island, *W. Austral., Austral.* **150** C1

Disappointment, Lake, *W. Austral., Austral.* **150** C2

Discovery Bay, *Austral.* **151** E4

Disko *see* Qeqertarsuaq, *island, Greenland, Den.* **30** B4

Dismal Mountains, *Antarctica* **167** B5

Divinhe, *Mozambique* **145** C4

Divo, *Côte d'Ivoire* **138** E2

Dixon Entrance, *Can., U.S.* **40** B1

Diyarbakır, *Turk.* **110** B3

Djado, *Niger* **139** C5

Djado, Plateau du, *Niger* **139** C4

Djambala, *Congo* **142** B2

Djelfa, *Alg.* **138** A3

Djibouti, *Af.* **126** C5

Djibouti, *Djibouti* **141** D4

Dnieper, *river, Ukr.* **79** C4

Dnipro, *river, Belarus, Ukr.* **77** C4

Dnipropetrovs'k, *Ukr.* **93** A6

Dnister, *river, Mold., Ukr.* **93** A4

Docampadó, Ensenada, *Col.* **68** B1

Dodecanese, *islands, Gr.* **93** E4

Dodoma, *Tanzania* **143** C5

Dogaicoring, *lake, China* **118** C2

Doha *see* Ad Dawḩah, *Qatar* **113** B4

Doi Inthanon, *peak, Thai.* **122** A1

Doka, *Sudan* **112** E1

Dokdo (Takeshima, Liancourt Rocks), *S. Kor.* **121** C4

Dolak, *island, Indonesia* **123** E6

Dolisie (Loubomo), *Congo* **142** C1

Dolo Bay, *Eth.* **141** E4

Dolonnur *see* Duolun, *China* **119** B5

Dolphin, Cape, *Falkland Is.* **73** E4

Dominica, *N. Amer.* **52** C4

Dominican Republic, *N. Amer.* **29** E5

Dominica Passage, *Dominica, Guadeloupe* **52** C4

Domuyo, Volcán, *Arg.* **72** C2

Domuyo Volcano, *Arg.* **59** E2

Don, *river, Russ.* **79** C5

Donau (Danube), *river, Aust., Ger.* **92** A2

Dondo, *Angola* **142** D2

Donegal Bay, *Ire.* **88** B2

Donets'k, *Ukr.* **93** A6

Dongara, *W. Austral., Austral.* **148** D1

Dongfang (Basuo), *China* **119** E4

Dongfeng, *China* **120** A1

Donghae, *S. Kor.* **120** C3

Dongola, *Sudan* **140** C2

Dongsha (Pratas Island), *China* **119** E5

Dongting Hu, *China* **119** D5

Donostia-San Sebastián, *Sp.* **89** E3

Dora, Lake, *W. Austral., Austral.* **150** C2

Dortmund, *Ger.* **91** E1

Dos Bahías, Cabo, *Arg.* **73** D3

Dosso, *Niger* **138** D3

Dossor, *Kaz.* **114** B2

Douala, *Cameroon* **142** A1

Double Island Point, *Qnsld., Austral.* **151** C6

Douglas, *Ariz., U.S.* **42** D2

Douglas, *U.K.* **88** B3

Douro, *river, Port.* **89** E2

Dover, *Del., U.S.* **43** C6

Dover, *U.K.* **88** C4

Dover, Strait of, *Eur.* **88** C4

Drâa, Hamada du, *Alg., Mor.* **138** B2

Dragon's Mouths, *strait, Trin. & Tobago, Venez.* **53** E4

Drakensberg, *mountains, Lesotho, S. Af., Swaziland* **129** F4

Drammen, *Nor.* **91** D2

Drau, *river, Aust.* **92** A2

Dráva, *river, Croatia, Hung.* **92** B3

Dresden, *Ger.* **91** F2

Drysdale, *river, W. Austral., Austral.* **150** A3

F

G

Ina-Jan

210

Ket-Kod

L

Las Heras, *Arg.* **73** E2
Lashio, *Myanmar* **117** C6
Lashkar Gah, *Afghan.* **116** A2
Las Palmas, *Sp.* **138** B1
La Spezia, *It.* **92** B1
Lassen Volcanic National Park, *U.S.* **46** B1
Las Tablas, *Pan.* **49** E6
Las Tunas, *Cuba* **50** C3
Las Vegas, *Nev., U.S.* **42** C2
Latady Island, *Antarctica* **166** B2
Latakia see Al Lādhiqīyah, *Syr.* **110** C2
La Tortuga, Isla, *Venez.* **53** F1
Latvia, *Eur.* **77** C4
Launceston, *Tas., Austral.* **148** E3
Lausanne, *Switz.* **92** A1
Laut, *island, Indonesia* **122** D3
Laval, *Fr.* **88** C3
La Vega, *Dom. Rep.* **51** D5
Laverton, *W. Austral., Austral.* **148** D2
Law Dome, *Antarctica* **167** D5
Lebanon, *Asia* **98** C2
Lebowakgomo, *S. Af.* **144** C3
Lecce, *It.* **92** D3
Leeuwin, Cape, *W. Austral., Austral.* **150** E1
Leeward Islands, *N. Amer.* **52** B2
Lefkoşa see Nicosia, *Cyprus* **93** E6
Lefkosia see Nicosia, *Cyprus* **93** E6
Lefroy, Lake, *W. Austral., Austral.* **150** D2
Legazpi, *Philippines* **123** B4
Le Havre, *Fr.* **88** C3
Leigh Creek, *S. Austral., Austral.* **149** D4
Leipzig, *Ger.* **91** F2
Le Lamentin, *Martinique, Fr.* **52** C4

Leleiwi Point, *Hawai'i, U.S.* **45** F5
Le Maire, Estrecho de, *Arg.* **73** F3
Léman, Lac, *Switz.* **89** D5
Le Mans, *Fr.* **88** C3
Lemesos (Limassol), *Cyprus* **93** E6
Lemnos see Límnos, *island, Gr.* **93** D4
Lenghu, *China* **118** C3
Lengua de Vaca Point, *Chile* **59** D2
Leninogorsk see Ridder, *Kaz.* **115** A6
Lenin Peak, *Kyrg., Taj.* **115** D5
Leninsk see Bayqongyr, *Kaz.* **114** C2
Lənkəran, *Azerb.* **111** B5
Leoben, *Aust.* **92** A2
León, *Mex.* **48** C3
León, *Nicar.* **49** D5
Leonora, *W. Austral., Austral.* **148** D2
Léopoldville see Kinshasa, *Dem. Rep. of the Congo* **142** C2
Lepsi, *Kaz.* **115** C5
Ler, *S. Sudan* **141** D2
Lerwick, *U.K.* **76** B2
Les Cayes, *Haiti* **51** D4
Lesotho, *Af.* **127** E4
Lesozavodsk, *Russ.* **121** A5
Les Saintes, *islands, Guadeloupe, Fr.* **52** C4
Lesser Antilles, *islands, N. Amer.* **52** B2
Lesser Sunda Islands, *Indonesia, Timor-Leste* **122** E3
Lésvos (Mytilíni), *island, Gr.* **93** D4
Leszno, *Pol.* **91** F2
Lethbridge, *Alberta, Can.* **40** D2
Lethem, *Guyana* **69** C5
Leticia, *Col.* **68** D3
Leveque, Cape, *W. Austral., Austral.* **150** B2

Lévrier, Baie du, *Mauritania* **138** C1
Lewis, Isle of, *U.K.* **78** B2
Lewis Range, *Austral.* **150** B3
Leyte, *island, Philippines* **123** B4
Lhasa, *China* **118** D2
Liancourt Rocks see Dokdo, *S. Kor.* **121** C4
Liaoyang, *China* **120** A1
Liaoyuan, *China* **119** B5
Liard, *river, N.W.T., Can.* **40** B2
Liberia, *Af.* **126** C1
Liberia, *Costa Rica* **49** E5
Libreville, *Gabon* **142** B1
Libya, *Af.* **126** B3
Libyan Desert, *Egypt, Lib., Sudan* **140** B3
Libyan Plateau (Aḑ Ḑiffah), *Egypt, Lib.* **140** A1
Lichinga, *Mozambique* **145** A4
Licosa, Punta, *It.* **92** D2
Liechtenstein, *Eur.* **76** D3
Liège, *Belg.* **88** C4
Liepāja, *Latv.* **91** D3
Ligurian Sea, *Eur.* **89** D5
Līhu'e, *Hawai'i, U.S.* **45** D1
Likasi, *Dem. Rep. of the Congo* **143** D4
Lillehammer, *Nor.* **76** B3
Lilongwe, *Malawi* **143** E5
Lima, *Peru* **70** C1
Limassol see Lemesos, *Cyprus* **93** E6
Limerick, *Ire.* **88** B2
Limfjorden, *bay, Den.* **91** D1
Limmen Bight, *N. Terr., Austral.* **151** A4
Límnos (Lemnos), *island, Gr.* **93** D4
Limoges, *Fr.* **89** D4
Limón, *Costa Rica* **49** E5
Limpopo, *river, Af.* **145** C4
Linares, *Sp.* **89** F3
Lincoln, *Nebr., U.S.* **42** C3
Lincoln Sea, *Can., Greenland* **165** D4
Lindesnes, *cape, Nor.* **91** D1

Maíz, Islas del, *Nicar.*
49 D5

Majorca *see* Mallorca,
island, Sp. **89** E4

Majuro, *Marshall Is.*
160 B3

Makarikari *see*
Makgadikgadi Pans,
Botswana **144** C3

Makassar (Ujungpandang),
Indonesia **123** E4

Makassar Strait, *Indonesia*
122 D3

Makgadikgadi Pans
(Makarikari), *Botswana*
144 C3

Makhachkala, *Russ.* **114** C1

Makhado, *S. Af.* **144** C3

Makkah (Mecca), *Saudi
Arabia* **112** C2

Makkovik, *Nfld. & Lab., Can.*
41 C5

Makokou, *Gabon* **142** B1

Mākoleʻa Point, *Hawaiʻi,
U.S.* **45** F4

Makurdi, *Nigeria* **139** E4

Malabo, *Eq. Guinea* **142** A1

Malacca, Strait of,
Indonesia, Malaysia
122 C1

Maladzyechna, *Belarus*
91 E4

Málaga, *Sp.* **89** F2

Malakal, *S. Sudan* **141** D2

Malang, *Indonesia* **122** E3

Malanje, *Angola* **142** D2

Malaspina Glacier, *Alas.,
U.S.* **44** B4

Malatya, *Turk.* **110** B3

Malawi, *Af.* **127** D4

Malawi, Lake (Lake
Nyasa), *Malawi,
Mozambique, Tanzania*
143 D5

Malay Peninsula, *Malaysia,
Thai.* **122** C1

Malaysia, *Asia* **99** E4

Maldive Islands, *Maldives*
100 E3

Maldives, *Asia* **98** E3

Male *see* Maale, *Maldives*
98 E3

Mali, *Af.* **126** C2

Malindi, *Kenya* **143** B6

Malin Head, *Ire.* **88** A2

Mallawi, *Egypt* **140** B2

Mallorca (Majorca), *island,
Sp.* **89** E4

Malmö, *Sw.* **91** E2

Malta, *Eur.* **76** E3

Malvinas, Islas *see* Falkland
Islands, *U.K.* **73** F3

Mammoth Cave National
Park, *U.S.* **47** C5

Mamoré, *river, Bol., Braz.*
70 C2

Man, *Côte d'Ivoire* **138** E2

Man, Isle of, *U.K.* **88** B3

Manado, *Indonesia* **123** D4

Managua, *Nicar.* **49** E5

Manakara, *Madagascar*
145 C6

Manama *see* Al Manāmah,
Bahrain **113** B4

Mananjary, *Madagascar*
145 C6

Manantiales, *Chile* **73** F2

Mānā Point, *Hawaiʻi, U.S.*
45 D1

Manaus, *Braz.* **69** D5

Manchester, *U.K.* **88** B3

Manchurian Plain, *China*
101 C5

Mand, *river, Iran* **111** D5

Mandalay, *Myanmar* **117** C6

Mandalgovĭ, *Mongolia*
119 B4

Mandimba, *Mozambique*
145 B4

Mangalore *see* Mangaluru,
India **116** D3

Mangaluru (Mangalore),
India **116** D3

Mangnai, *China* **118** C3

Mangole, *island, Indonesia*
123 D4

Manicoré, *Braz.* **69** E4

Manicouagan, Réservoir,
Que., Can. **41** D5

Manīfah, *Saudi Arabia*
113 B4

Manifold, Cape, *Qnsld.,
Austral.* **151** C6

Maniitsoq (Sukkertoppen),
Greenland, Den. **165** E5

Manila, *Philippines* **123** B4

Manily, *Russ.* **95** B6

Manipa, Selat, *Indonesia*
123 D5

Manitoba, *province, Can.*
40 D3

Manitoba, Lake, *Man., Can.*
40 D3

Manizales, *Col.* **68** B2

Manja, *Madagascar* **145** C5

Manjimup, *W. Austral.,
Austral.* **148** E2

Mannar, Gulf of, *India, Sri
Lanka* **117** E4

Manono, *Dem. Rep. of the
Congo* **143** C4

Manpʼo, *N. Kor.* **120** B2

Mansa, *Zambia* **143** D4

Mansel Island, *Nunavut,
Can.* **41** C4

Manta, *Ecua.* **68** C1

Manyara, Lake, *Tanzania*
143 B5

Manzanillo, *Cuba* **50** D3

Manzanillo, *Mex.* **48** C2

Mao, *Chad* **139** D5

Mapai, *Mozambique*
145 C4

Mapimí, Bolsón de, *Mex.*
31 D2

Mapuera, *river, Braz.*
69 C5

Maputo, *Mozambique*
145 D4

Maputo, Baía de,
Mozambique **145** D4

Maqat, *Kaz.* **114** B2

Marabá, *Braz.* **71** B4

Maracaibo, *Venez.* **68** A2

Maracaibo, Lago de, *Venez.*
68 A2

Maracaibo, Lake, *Venez.*
58 A2

Maracay, *Venez.* **68** A3

Maradi, *Niger* **139** D4

Marāgheh, *Iran* **111** B4

Marahuaca, Cerro, *Venez.*
69 C4

Marajó, Baía de, *Braz.*
71 A5

Marajó, Ilha de, *Braz.*
71 A4

Marajó Island, *Braz.* **58** B4

Maralal, *Kenya* **143** A6

Marambio, *research station, Antarctica* **166** Al

Marañón, *river, Peru*
68 D2

Marble Bar, *W. Austral., Austral.* **148** B2

Mar Chiquita, Laguna, *Arg.*
72 B3

Mar del Plata, *Arg.* **72** C4

Mardin, *Turk.* **110** B3

Margarita, Isla de, *Venez.*
69 A4

Margarita Island, *Venez.*
58 A2

Margherita Peak, *Dem. Rep. of the Congo, Uganda*
143 B4

Mariana Islands, *U.S.*
101 C6

Marías, Islas, *Mex.* **48** C2

Marías Islands, *Mex.* **31** E2

Maribor, *Slov.* **92** B2

Maridi, *S. Sudan* **141** E2

Marie Byrd Land, *Antarctica* **166** D3

Marie-Galante, *island, Guadeloupe, Fr.* **52** B4

Mariental, *Namibia* **144** C2

Marigot, *St. Martin, Fr.*
52 A3

Marir, Gezâir (Mirear), *Egypt* **112** Cl

Mariscal Estigarribia, *Parag.* **70** D3

Mariupol', *Ukr.* **93** A6

Marka (Merca), *Somalia*
141 E4

Markala, *Mali* **138** D2

Marmagao, *India* **116** D3

Marmara, Sea of, *Turk.*
79 E4

Marmara Denizi, *Turk.*
110 Al

Maroantsetra, *Madagascar*
145 B6

Maromokotro, *peak, Madagascar* **145** B6

Maroua, *Cameroon* **139** D5

Marovoay, *Madagascar*
145 B6

Marquesas Islands, *Fr. Polynesia, Fr.* **161** C6

Marra, Jebel, *Sudan* **141** Dl

Marrakech, *Mor.* **138** A2

Marree, *S. Austral., Austral.*
149 C4

Marromeu, *Mozambique*
145 B4

Marrupa, *Mozambique*
145 A4

Marsabit, *Kenya* **143** A6

Marseille, *Fr.* **89** D5

Marshall Islands, *Oceania*
160 B3

Martaban, Gulf of, *Myanmar* **117** D6

Martapura, *Indonesia*
122 D3

Martinique, *possession, Fr.*
52 C4

Martinique Passage, *Dominica, Martinique*
52 C4

Mary, *Turkm.* **114** E3

Maryborough, *Qnsld., Austral.* **149** C6

Maryland, *state, U.S.*
43 C6

Mary's Harbour, *Nfld. & Lab., Can.* **41** C6

Marzo, Cabo, *Col.* **68** Bl

Marzūq, Şaḩrā', *Lib.*
139 B5

Masai Steppe, *Tanzania*
143 C5

Masan, *S. Kor.* **120** D2

Masasi, *Tanzania* **143** D6

Masbate, *island, Philippines*
123 B4

Masbate, *Philippines*
123 B4

Maseru, *Lesotho* **144** D3

Mashhad, *Iran* **111** B6

Maşira *see* Maşīrah, Jazīrat, *Oman* **113** C6

Masira, Gulf of, *Oman*
100 D2

Maşīrah, Jazīrat (Masira), *Oman* **113** C6

Masjed Soleymān, *Iran*
111 C5

Masqaṭ (Muscat), *Oman*
113 C6

Massachusetts, *state, U.S.*
43 B6

Massangena, *Mozambique*
145 C4

Massawa *see* Mits'iwa, *Eritrea* **140** C4

Massif Central, *mountains, Fr.* **89** D4

Masvingo, *Zimb.* **144** C3

Matadi, *Dem. Rep. of the Congo* **142** Cl

Matamoros, *Mex.* **48** B3

Matanzas, *Cuba* **50** C2

Matarani, *Peru* **70** D2

Matay, *Kaz.* **115** C5

Mathew Town, *Bahamas*
51 C4

Mato Grosso, Planalto do, *Braz.* **70** C3

Mato Grosso Plateau, *Braz.*
58 C3

Mátra, *peak, Hung.* **92** A3

Maṭraḥ, *Oman* **113** C5

Matsue, *Japan* **121** D4

Matsumoto, *Japan* **121** D5

Matsuyama, *Japan* **121** D4

Matterhorn, *peak, It., Switz.*
89 D5

Maturín, *Venez.* **69** A4

Mauá, *Mozambique*
145 A4

Maués-Açu, *river, Braz.*
70 B3

Maui, *island, Hawai'i, U.S.*
45 E4

Maun, *Botswana* **144** C2

Mauna Kea, *peak, Hawai'i, U.S.* **45** F5

Mauna Loa, *peak, Hawai'i, U.S.* **45** F5

Mexico, Gulf of, *N. Amer.*
31 E3

Mexico City, *Mex.* 48 C3

Meymaneh, *Afghan.* 114 E3

Mezen', *Russ.* 77 A5

Mezen' Bay, *Russ.* 79 A5

Miami, *Fla., U.S.* 43 E5

Miami Beach, *Fla., U.S.*
50 B2

Michigan, *state, U.S.* 43 B5

Michigan, Lake, *U.S.*
43 B4

Micronesia, *islands, Pacific
Oc.* 160 B2

Middelburg, *S. Af.* 144 E3

Middle Andaman, *island,
India* 117 D5

Middlesbrough, *U.K.*
88 B3

Midway Islands, *possession,
U.S.* 161 A4

Miho Wan, *Japan* 121 D4

Mikhaiylovskiy, *Russ.*
115 A5

Mikhaylova, *Russ.* 94 B3

Milan *see* Milano, *It.* 92 B1

Milano (Milan), *It.* 92 B1

Mildura, *Vic., Austral.*
149 E4

Miles City, *Mont., U.S.*
42 B3

Mililani Town, *Hawai'i, U.S.*
45 D3

Miller Range, *Antarctica*
167 D4

Mill Island, *Antarctica*
167 D6

Milwaukee, *Wis., U.S.*
43 B4

Mīnāb, *Iran* 111 E6

Minatitlán, *Mex.* 49 D4

Mindanao, *island,
Philippines* 123 C4

Mindoro, *island, Philippines*
123 B4

Minfeng (Niya), *China*
118 C2

Mingan, *Que., Can.* 41 D5

Minigwal, Lake, *W. Austral.,
Austral.* 150 D2

Minneapolis, *Minn., U.S.*
43 B4

Minnesota, *state, U.S.*
43 B4

Minorca *see* Menorca,
island, Sp. 89 E4

Minot, *N. Dak., U.S.* 42 A3

Minsk, *Belarus* 91 E4

Minxian, *China* 119 C4

Mirear *see* Marir, Gezâir,
Egypt 112 C1

Miri, *Malaysia* 122 C3

Mirik *see* Timiris, Cap,
Mauritania 138 C1

Mirim, Lagoa, *Braz.* 72 B5

Mīrjāveh, *Iran* 113 A6

Mirnyy, *research station,
Antarctica* 167 C6

Mirnyy, *Russ.* 95 C4

Miskitos, Cayos, *Nicar.*
49 D5

Mişrātah, *Lib.* 139 A5

Mississippi, *river, U.S.*
43 D4

Mississippi, *state, U.S.*
43 D4

Mississippi River Delta,
U.S. 31 D3

Missoula, *Mont., U.S.*
42 B2

Missouri, *river, U.S.* 43 C4

Missouri, *state, U.S.* 43 C4

Mitau *see* Jelgava, *Latv.*
91 D4

Mitchell, *river, Qnsld.,
Austral.* 151 B5

Mitilíni, *Gr.* 93 D4

Mito, *Japan* 121 C6

Mits'iwa (Massawa),
Eritrea 140 C4

Mitú, *Col.* 68 C3

Mitumba Mountains,
Dem. Rep. of the Congo
143 D4

Miyako Jima, *Japan* 119 D6

Miyazaki, *Japan* 121 E4

Moa, *island, Indonesia*
123 E5

Moanda, *Gabon* 142 B1

Mobile, *Ala., U.S.* 43 D4

Moçambique, *Mozambique*
145 B5

Mochudi, *Botswana* 144 C3

Mocuba, *Mozambique*
145 B4

Modica, *It.* 92 E2

Mogadishu *see* Muqdisho,
Somalia 141 E5

Mogok, *Myanmar* 117 C6

Mohéli *see* Mwali, *island,
Comoros* 145 A5

Mo i Rana, *Nor.* 90 B2

Mojave Desert, *Calif., U.S.*
42 C1

Mokpo, *S. Kor.* 120 E2

Moldova, *Eur.* 77 D4

Molepolole, *Botswana*
144 C3

Moloa'a Bay, *Hawai'i, U.S.*
45 D2

Moloka'i, *island, Hawai'i,
U.S.* 45 E3

Moluccas, *islands, Indonesia*
123 D5

Molucca Sea, *Indonesia*
123 D4

Moma, *Mozambique*
145 B5

Mombasa, *Kenya* 143 C6

Mona, Isla, *P.R., U.S.* 51 D6

Monaco, *Eur.* 76 D3

Mona Passage, *Dom. Rep.,
P.R.* 51 D6

Monchegorsk, *Russ.*
90 A4

Monclova, *Mex.* 48 B3

Mondah, Baie de, *Eq.
Guinea, Gabon* 142 B1

Mongers Lake, *W. Austral.,
Austral.* 150 D1

Mongo, *Chad* 139 D5

Mongolia, *Asia* 99 C4

Mongolian Plateau, *China,
Mongolia* 101 C4

Mongu, *Zambia* 142 E3

Monrovia, *Liberia* 138 E2

Montana, *state, U.S.* 42 B2

Monte Alegre, *Braz.* 71 A4

Monte Cristi, *Dom. Rep.*
51 D5

Montego Bay, *Jam.* 50 D3

Nao, Cabo de la, *Sp.* **89** F4

Nã Pali Coast, *Hawai'i, U.S.* **45** DI

Naples *see* Napoli, *It.* **92** C2

Napo, *river, Ecua., Peru* **68** D2

Napoli (Naples), *It.* **92** C2

Napoli, Golfo di, *It.* **92** C2

Nara, *Mali* **138** D2

Narayanganj, *Bangladesh* **117** C5

Narbonne, *Fr.* **89** E4

Nares Strait, *Can., Greenland* **41** A6

Naricual, *Venez.* **53** FI

Narinda, Baie de, *Madagascar* **145** B6

Narodnaya, Gora, *Russ.* **79** A5

Narsarsuaq, *Greenland, Den.* **165** E6

Narva, *Est.* **91** D4

Narvik, *Nor.* **90** A3

Nar'yan Mar, *Russ.* **165** A5

Naryn, *Kyrg.* **115** D5

Nasca, *Peru* **70** CI

Nashville, *Tenn., U.S.* **43** C5

Nasiriyah *see* An Nãşirīyah, *Iraq* **111** D4

Nassau, *Bahamas* **50** B3

Nasser, Lake, *Egypt* **140** B2

Nata, *Botswana* **144** C3

Natal, *Braz.* **71** B6

Natara, *Russ.* **95** C4

Natuna Besar, Kepulauan, *Indonesia* **122** C2

Natuna Selatan, Kepulauan, *Indonesia* **122** C2

Naturaliste, Cape, *W. Austral., Austral.* **150** EI

Nauru, *Oceania* **160** B3

Navarin, Mys, *Russ.* **95** A6

Navarino, Isla, *Chile* **73** F2

Navoiy, *Uzb.* **114** D3

Navojoa, *Mex.* **48** B2

Nāwiliwili Bay, *Hawai'i, U.S.* **45** DI

Nayoro, *Japan* **121** B6

Nay Pyi Taw, *Myanmar* **117** C6

Nayramadln Orgil (Youyi Feng), *China, Mongolia, Russ.* **118** A2

Nazrēt, *Eth.* **141** D3

N'dalatando, *Angola* **144** AI

N'Djamena, *Chad* **139** D5

Ndola, *Zambia* **143** D4

Neale, Lake, *N. Terr., Austral.* **150** C3

Near Islands, *Alas., U.S.* **44** C3

Nebitdag *see* Balkanabat, *Turkm.* **114** D2

Neblina, Pico da, *Braz., Venez.* **68** C3

Nebraska, *state, U.S.* **42** B3

Necochea, *Arg.* **72** C4

Neftçala, *Azerb.* **77** D6

Negomano, *Mozambique* **145** A4

Negro, *river, Arg.* **72** C3

Negro, *river, Braz.* **69** D4

Negros, *island, Philippines* **123** C4

Nehbandãn, *Iran* **113** A6

Nehe, *China* **119** A5

Neiba, *Dom. Rep.* **51** D5

Neiba, Bahía de, *Dom. Rep.* **51** D5

Neisse, *river, Ger., Pol.* **91** F2

Neiva, *Col.* **68** C2

Nelidovo, *Russ.* **91** D5

Nellore, *India* **117** D4

Nelson, *river, Man., Can.* **40** C3

Nelson, Cape, *Vic., Austral.* **151** E4

Nelspruit, *S. Af.* **145** C4

Néma, *Mauritania* **138** D2

Neman, *river, Belarus* **91** E4

Nepal, *Asia* **98** D3

Neriquinha, *Angola* **144** B2

Netherlands, *Eur.* **76** C2

Netzahualcóyotl, *Mex.* **48** C3

Neumayer, *research station, Antarctica* **166** A3

Neuquén, *river, Arg.* **72** C2

Neuquén, *Arg.* **72** C2

Nevada, *state, U.S.* **42** BI

Nevada, Sierra, *U.S.* **42** CI

Nevers, *Fr.* **89** D4

Nevis, Ben, *peak, U.K.* **88** A3

New Amsterdam, *Guyana* **69** B5

New Brunswick, *province, Can.* **41** E5

New Caledonia, *possession, Fr.* **160** D3

Newcastle, *N.S.W., Austral.* **149** E6

Newcastle, *U.K.* **88** B3

Newcastle Waters, *N. Terr., Austral.* **149** B4

New Delhi, *India* **116** B3

Newenham, Cape, *Alas., U.S.* **44** C2

Newfoundland, Island of, *Nfld. & Lab., Can.* **41** D6

Newfoundland and Labrador, *province, Can.* **41** C5

New Guinea, *island, Indonesia, P.N.G.* **160** CI

New Hampshire, *state, U.S.* **43** B6

New Jersey, *state, U.S.* **43** B6

Newman, *W. Austral., Austral.* **148** C2

Newman Island, *Antarctica* **166** D3

New Mexico, *state, U.S.* **42** D2

New Orleans, *La., U.S.* **43** D4

New Providence, *island, Bahamas* **50** B2

New Schwabenland, *region, Antarctica* **166** A4

New Siberian Islands, *Russ.* **101** A4

Ogilvie Mountains, *Yukon, Can.* **44** B4

Ogoki, *Ont., Can.* **41** D4

Ohio, *river, U.S.* **43** C5

Ohio, *state, U.S.* **43** C5

Ohridsko Jezero, *Alban., Maced.* **92** C3

Oiapoque *see* Oyapock, *river, Braz.* **69** C6

Oiapoque, *Braz.* **69** B6

Ōita, *Japan* **121** D4

Okahandja, *Namibia* **144** C2

Okavango, *river, Angola, Botswana, Namibia* **144** B2

Okavango Delta, *Botswana* **144** B2

Okeechobee, Lake, *Fla., U.S.* **43** E5

Okha, *Russ.* **95** C6

Okhotsk, *Russ.* **95** C5

Okhotsk, Sea of, *Japan, Russ.* **101** B5

Okinawa, *island, Japan* **119** D6

Oki Shotō, *Japan* **121** D4

Oklahoma, *state, U.S.* **42** D3

Oklahoma City, *Okla., U.S.* **42** C3

Okp'yŏng, *N. Kor.* **120** B2

Oktyabr'sk, *Kaz.* **114** A2

Okushiri Tō, *Japan* **121** B5

Öland, *island, Sw.* **91** D2

Olavarría, *Arg.* **72** C4

Olbia, *It.* **92** C1

Old Bahama Channel, *Cuba* **50** C3

Old Crow, *Yukon, Can.* **40** A2

Olekminsk, *Russ.* **95** C5

Olenekskiy Zaliv, *Russ.* **164** A3

Olga, Mount (Kata Tjuṯa), *N. Terr., Austral.* **150** C3

Ölgiy, *Mongolia* **118** A3

Ólimbos (Olympus), *peak, Gr.* **92** D3

Olomouc, *Czech Rep.* **92** A3

Olsztyn, *Pol.* **91** E3

Olt, *river, Rom.* **93** B4

Olympia, *Wash., U.S.* **42** A1

Olympic National Park, *U.S.* **46** A1

Olympus *see* Ólimbos, *peak, Gr.* **92** D3

Olyutorskiy, Mys, *Russ.* **95** A6

Omaha, *Nebr., U.S.* **42** C3

Oman, *Asia* **98** D2

Oman, Gulf of, *Iran, Oman, U.A.E.* **113** B5

Omaruru, *Namibia* **144** C1

Omdurman, *Sudan* **140** C2

Ometepec, *Mex.* **48** D3

Omolon, *Russ.* **95** B5

Omsk, *Russ.* **94** D3

Ondjiva, *Angola* **142** E2

Onega, *Russ.* **90** B5

Onega, Lake, *Russ.* **79** B4

Onega Bay, *Russ.* **79** A4

Onezhskaya Guba, *Russ.* **90** B5

Onezhskoye Ozero, *Russ.* **90** C5

Ongjin, *N. Kor.* **120** C1

Onslow, *W. Austral., Austral.* **148** C1

Ontario, *province, Can.* **40** D4

Ontario, Lake, *Can., U.S.* **43** B5

Oodnadatta, *S. Austral., Austral.* **149** C4

'Ōpana Point, *Hawai'i, U.S.* **45** E4

Oporto *see* Porto, *Port.* **89** E2

Opuwo, *Namibia* **144** B1

Oradea, *Rom.* **92** B3

Oral, *Kaz.* **114** A2

Oran, *Alg.* **138** A3

Orange, *N.S.W., Austral.* **149** D5

Orange (Oranje), *river, Lesotho, Namibia, S. Af.* **144** D2

Orange, Cabo, *Braz.* **69** B6

Orange Walk, *Belize* **49** C5

Oranje *see* Orange, *river, Lesotho, Namibia, S. Af.* **144** D2

Oranjestad, *Aruba, Neth.* **53** D1

Oranjestad, *St. Eustatius, Neth.* **52** A3

Orcadas, *research station, Antarctica* **166** A2

Ord, *river, W. Austral., Austral.* **150** A3

Ord, Mount, *W. Austral., Austral.* **150** B2

Örebro, *Sw.* **91** D2

Oregon, *state, U.S.* **42** B1

Orel, *Russ.* **77** C5

Orenburg, *Russ.* **77** C6

Oriental, Cordillera, *S. Amer.* **68** D1

Orinoco, *river, Col., Venez.* **69** B4

Orinoco River Delta, *Venez.* **58** A3

Oristano, *It.* **92** D1

Orizaba, Pico de, *Mex.* **31** E3

Orkney Islands, *U.K.* **88** A3

Orlando, *Fla., U.S.* **43** D5

Orléans, *Fr.* **88** C4

Örnsköldsvik, *Sw.* **76** B3

Orsha, *Belarus* **91** E5

Orsk, *Russ.* **77** B6

Orto Surt, *Russ.* **95** C5

Orūmīyeh (Urmia), *Iran* **111** B4

Orūmīyeh, Daryācheh-ye, *Iran* **111** B4

Oruro, *Bol.* **70** D2

Ōsaka, *Japan* **121** D5

Ösel *see* Saaremaa, *island, Est.* **91** D3

Osh, *Kyrg.* **115** D5

Oshakati, *Namibia* **144** B1

Osijek, *Croatia* **92** B3

Öskemen (Ust' Kamenogorsk), *Kaz.* **115** B6

Oslo, *Nor.* **90** C2

Osorno, *Chile* **72** C1

Port Étienne *see*
Nouadhibou, *Mauritania*
138 CI

Port-Gentil, *Gabon* **142** BI

Port Harcourt, *Nigeria*
139 E4

Port Hedland, *W. Austral.,*
Austral. **148** B2

Port Herald *see* Nsanje,
Malawi **145** B4

Port Láirge *see* Waterford,
Ire. **88** B2

Portland, *Me., U.S.* **43** B6

Portland, *Oreg., U.S.* **42** AI

Portland Point, *Jam.*
50 E3

Port Lincoln, *S. Austral.,*
Austral. **149** E4

Port-Louis, *Guadeloupe, Fr.*
52 B4

Port Lyautey *see* Kenitra,
Mor. **138** A2

Port Macquarie, *N.S.W.,*
Austral. **149** D6

Port Moresby, *P.N.G.*
160 C2

Port Nolloth, *S. Af.* **144** B4

Porto (Oporto), *Port.*
89 E2

Porto Alegre, *Braz.* **72** B5

Porto Amboim, *Angola*
142 D2

Port of Spain, *Trin. &*
Tobago **53** F4

Porto Nacional, *Braz.*
71 C5

Porto-Novo, *Benin* **138** E3

Porto-Vecchio, *Fr.* **92** CI

Porto Velho, *Braz.* **69** E4

Portoviejo, *Ecua.* **68** DI

Port Phillip Bay, *Vic.,*
Austral. **151** E4

Port Pirie, *S. Austral.,*
Austral. **149** D4

Port Said *see* Bûr Sa'îd,
Egypt **112** AI

Portsmouth, *Dominica*
52 C4

Port Sudan, *Sudan* **140** C3

Portugal, *Eur.* **76** DI

Port-Vila, *Vanuatu* **160** C3

Pòtoprens (Port-au-Prince),
Haiti **51** D4

Posadas, *Arg.* **72** A5

Posadowsky Bay, *Antarctica*
167 C6

Postmasburg, *S. Af.*
144 D3

Potiskum, *Nigeria* **139** D4

Potosí, *Bol.* **70** D2

Poza Rica, *Mex.* **48** C3

Poznań, *Pol.* **91** E2

Prague *see* Praha, *Czech*
Rep. **92** A2

Praha (Prague), *Czech Rep.*
92 A2

Prainha, *Braz.* **69** E4

Pratas Island *see* Dongsha,
China **119** E5

Preparis Island, *Myanmar*
117 D5

Pressburg *see* Bratislava,
Slovakia **92** A3

Pretoria (Tshwane), *S. Af.*
144 D3

Pribilof Islands, *U.S.* **31** F2

Prieska, *S. Af.* **144** D2

Prilep, *Maced.* **92** C3

Prince Albert, *Sask., Can.*
40 D2

Prince Charles Mountains,
Antarctica **167** B5

Prince Edward Island,
province, Can. **41** D6

Prince of Wales Island,
Nunavut, Can. **40** A3

Prince of Wales Island,
Alas., U.S. **44** C5

Prince Patrick Island, *Can.*
30 A2

Prince Rupert, *B.C., Can.*
40 CI

Princess Charlotte Bay,
Qnsld., Austral. **151** A5

Prince William Sound, *Alas.,*
U.S. **44** B3

Príncipe da Beira, *Braz.*
70 C3

Pristina *see* Prishtinë, *Kos.*
92 C3

Prishtinë (Pristina), *Kos.*
92 C3

Prizren, *Kos.* **92** C3

Progress 2, *research*
station, Antarctica **167** B5

Propriá, *Braz.* **71** C6

Prorva, *Kaz.* **114** C2

Providence, *R.I., U.S.*
43 B6

Provideniya, *Russ.* **164** CI

Provo, *Utah, U.S.* **42** C2

Prudhoe Bay, *Alas., U.S.*
44 A3

Prudhoe Bay, *Alas., U.S.*
44 A3

Prut, *river, Mold., Rom.*
93 B5

Prydz Bay, *Antarctica*
167 B5

Prypyats', *river, Belarus*
91 E4

Pskov, *Russ.* **91** D4

Pucallpa, *Peru* **68** E2

Puducherry (Pondicherry),
India **117** D4

Puebla, *Mex.* **29** E2

Pueblo Nuevo, *Venez.*
53 DI

Puerto Aysén, *Chile* **73** DI

Puerto Ángel, *Mex.* **48** D3

Puerto Ayacucho, *Venez.*
68 B3

Puerto Cabello, *Venez.*
68 A3

Puerto Cabezas, *Nicar.*
49 D5

Puerto Coig, *Arg.* **73** E2

Puerto de Hierro, *Venez.*
53 F3

Puerto Deseado, *Arg.*
73 E3

Puerto La Cruz, *Venez.*
53 FI

Puerto Madryn, *Arg.*
73 D3

Puerto Maldonado, *Peru*
70 C2

Puerto Montt, *Chile* **73** DI

Puerto Natales, *Chile*
73 F2

Puerto Píritu, *Venez.* **53** FI

Puerto Plata, *Dom. Rep.*
51 D5

Q

Rig-Rze

S

Saaremaa (Ösel), *island, Est.* **91** D3

Saba, *possession, Neth.* **52** A2

Sabana, Archipiélago de, *Cuba* **50** B2

Sabhā, *Lib.* **139** B5

Sable, Cape, *Fla., U.S.* **50** B2

Sable, Cape, *N.S., Can.* **41** E6

Sable Island, *N.S., Can.* **41** E6

Şabyā, *Saudi Arabia* **112** D2

Sabzevār, *Iran* **111** B6

Sachs Harbour, *N.W.T., Can.* **40** A2

Sacramento, *Calif., U.S.* **42** C1

Şa'dah, *Yemen* **112** D3

Sado, *island, Japan* **121** C5

Şafājah, *mountains, Saudi Arabia* **110** E3

Şafājah, *Saudi Arabia* **112** B2

Safi, *Mor.* **138** A2

Saga, *Japan* **120** E3

Sagami Nada, *Japan* **121** D6

Saghyz, *Kaz.* **114** B2

Saguaro National Park, *U.S.* **46** D2

Saguenay, *Que., Can.* **41** D5

Sahamalaza, Baie de (Port Radama), *Madagascar* **145** B6

Sahara, *desert, Af.* **128** B2

Sahel, *region, Af.* **128** C2

Sa'īdābād *see* Sīrjān, *Iran* **113** A5

Saigon *see* Thành Phố Hồ Chí Minh, *Vietnam* **122** B2

Saint André, Cap, *Madagascar* **145** B5

Saint Ann's Bay, *Jam.* **50** D3

Saint Augustin, Baie de, *Madagascar* **145** C5

Saint Augustine, *Fla., U.S.* **43** D5

Saint Barthélemy (Saint Barts), *possession, Fr.* **52** A3

Saint Barts *see* Saint Barthélemy, *possession, Fr.* **52** A3

Saint-Brieuc, *Fr.* **88** C3

Saint Croix, *island, U.S. Virgin Is., U.S.* **52** A1

Saint-Dizier, *Fr.* **88** C4

Saint Elias Mountains, *Can., U.S.* **44** B4

Sainte Marie, Cap, *Madagascar* **145** D5

Sainte Marie, Cape, *Madagascar* **129** E5

Sainte Marie, Nosy, *Madagascar* **145** B6

Sainte-Marie, *Martinique, Fr.* **52** C4

Saint Eustatius (Statia), *possession, Neth.* **52** A3

Saint Francis Bay, *S. Af.* **144** E3

Saint George's Channel, *Eur.* **88** B2

Saint George's, *Grenada* **53** E4

Saint Helena, *island, U.K.* **129** E1

Saint Helena Bay, *S. Af.* **144** E2

Saint John, *N.B., Can.* **41** E5

Saint John, *island, U.S. Virgin Is., U.S.* **52** A1

Saint John's, *Antigua & Barbuda* **52** B4

Saint John's, *Nfld. & Lab., Can.* **41** D6

Saint Joseph, *Trin. & Tobago* **53** F4

Saint Kitts and Nevis, *N. Amer.* **52** B3

Saint-Laurent du Maroni, *Fr. Guiana, Fr.* **69** B6

Saint Lawrence, *river, Can., U.S.* **41** D5

Saint Lawrence, Gulf of, *Can.* **41** D6

Saint Lawrence Island, *Alas., U.S.* **44** B1

Saint Louis, *Mo., U.S.* **43** C4

Saint-Louis, *Senegal* **138** D1

Saint Lucia, *N. Amer.* **53** D4

Saint Lucia Channel, *Martinique, St. Lucia* **53** D4

Saint Maarten (Saint Martin), *possession, Fr., Neth.* **52** A2

Saint Martin *see* Saint Maarten, *possession, Fr., Neth.* **52** A3

Saint-Mathieu, Point, *Fr.* **78** C2

Saint-Mathieu, Pointe de, *Fr.* **88** C2

Saint Matthew Island, *Alas., U.S.* **44** B1

Saint Paul, *Minn., U.S.* **43** B4

Saint Petersburg, *Fla., U.S.* **43** E5

Saint Petersburg *see* Sankt-Peterburg, *Russ.* **94** B1

Saint-Pierre and Miquelon Islands, *possession, Fr.* **41** D6

Saint Pölten, *Aust.* **92** A2

Saint Thomas, *island, U.S. Virgin Is., U.S.* **52** A1

Saint Vincent, *island, St. Vincent & the Grenadines* **53** D4

Saint Vincent, Cape, *Port.* **78** E1

Saint Vincent, Gulf, *S. Austral., Austral.* **151** E4

Saint Vincent and the Grenadines, *N. Amer.* **53** E4

Sai-San

T

Tim-Tra

Timişoara, *Rom.* **92** B3

Timmiarmiut, *Greenland, Den.* **165** E6

Timmins, *Ont., Can.* **41** E4

Timor, *island, Indonesia, Timor-Leste* **123** E4

Timor-Leste (East Timor), *Asia* **99** E5

Timor Sea, *Austral., Indonesia, Timor-Leste* **160** Cl

Tindouf, *Alg.* **138** B2

Tinrhert, Hamada de, *Alg., Lib.* **139** B4

Ti-n-Zaouâtene (Fort Pierre Bordes), *Alg.* **138** C3

Tirana see Tiranë, *Alban.* **92** C3

Tiranë (Tirana), *Alban.* **92** C3

Tiraspol, *Mold.* **93** B5

Tiruchchirappalli, *India* **117** E4

Tirunelveli, *India* **116** E3

Tisa, *river, Serb.* **92** B3

Tisza, *river, Hung.* **92** B3

Titan Dome, *Antarctica* **167** C4

Titicaca, Lago, *Bol., Peru* **70** C2

Titicaca, Lake, *Bol., Peru* **58** C2

Tizimín, *Mex.* **49** C5

Tizi Ouzou, *Alg.* **89** F4

Toamasina, *Madagascar* **145** B6

Tobago, *island, Trin. & Tobago* **53** F5

Tobi, *island, Palau* **123** C5

Tobi Shima, *Japan* **121** C6

Tobol, *river, Russ.* **114** A3

Tobol'sk, *Russ.* **76** A6

Tobruk see Ţubruq, *Lib.* **139** A6

Tobyl, *river, Kaz.* **114** A3

Tocantins, *river, Braz.* **71** A4

Toco, *Trin. & Tobago* **53** F4

Tocopilla, *Chile* **70** D2

Todos os Santos Bay, *Braz.* **58** C5

Togiak, *Alas., U.S.* **44** C2

Togo, *Af.* **126** C2

Tokar, *Sudan* **140** C3

Tokara Kaikyō (Colnett Strait), *Japan* **121** E4

Tokara Rettō, *Japan* **121** E4

Tŏkch'ŏn, *N. Kor.* **120** B2

Tokelau, *possession, N.Z.* **161** C4

Tokuno Shima, *Japan* **119** D6

Tokushima, *Japan* **121** D5

Tōkyō, *Japan* **121** D6

Tōkyō Wan, *Japan* **121** D6

Tôlan''aro, *Madagascar* **145** D6

Toliara, *Madagascar* **145** C5

Tolitoli, *Indonesia* **123** D4

Tolo, Teluk, *Indonesia* **123** D4

Tomakomai, *Japan* **121** B6

Tombouctou (Timbuktu), *Mali* **138** D3

Tombua, *Angola* **142** E1

Tomini, Teluk, *Indonesia* **123** D4

Tom Price, *W. Austral., Austral.* **148** C1

Tomsk, *Russ.* **94** D3

Tonalá, *Mex.* **49** D4

Tonga, *Oceania* **161** C4

Tonga Islands, *Tonga* **161** D4

Tonghua, *China* **120** Al

Tongliao, *China* **119** B5

Tongtian, *river, China* **118** C3

Tongue of the Ocean, *bay, Bahamas* **50** B3

Tonkin, Gulf of, *China, Vietnam* **119** E4

Tonle Sap, *lake, Cambodia* **122** B2

Toowoomba, *Qnsld., Austral.* **149** D6

Topeka, *Kans., U.S.* **43** C4

Torghay, *Kaz.* **114** B3

Torino, *It.* **89** D5

Tori Shima, *Japan* **121** E6

Torneälven, *river, Sw.* **90** B3

Torniojoki, *river, Fin.* **90** B3

Toro, Cerro del, *Arg., Chile* **72** A2

Toronto, *Ont., Can.* **41** E4

Toros Dağları, *Turk.* **110** B2

Torre de Cerredo, *peak, Sp.* **89** D2

Torrens, Lake, *S. Austral., Austral.* **151** D4

Torreón, *Mex.* **48** B2

Torres Strait, *Austral., P.N.G.* **151** A5

Tórshavn, *Faroe Is.* **165** C6

Tortola, *British Virgin Is., U.K.* **52** Al

Tortosa, *Sp.* **89** E4

Tortosa, Cap de, *Sp.* **76** D2

Tortosa, Cape, *Sp.* **78** D2

Tortue, Île de la, *Haiti* **51** D4

Ţorūd, *Iran* **111** B6

Tosa Wan, *Japan* **121** D5

Toshkent (Tashkent), *Uzb.* **115** D4

Tottori, *Japan* **121** D5

Toubkal, Jebel, *Mor.* **138** A2

Touggourt, *Alg.* **139** A4

Toulon, *Fr.* **89** D5

Toulouse, *Fr.* **89** D4

Toummo, *Lib.* **139** C5

Tours, *Fr.* **88** C3

Townsville, *Qnsld., Austral.* **149** B5

Toyama, *Japan* **121** C5

Toyama Wan, *Japan* **121** C5

Toyohashi, *Japan* **121** D5

Trâblous (Tripoli), *Lebanon* **111** C4

Trabzon, *Turk.* **110** A3

Trangan, *island, Indonesia* **123** E5

Transantarctic Mountains, *Antarctica* **166** B3

U

V

Virgin Gorda, *British Virgin Is., U.K.* **52** A2

Virginia, *state, U.S.* **43** C5

Virgin Islands, *Caribbean Sea* **31** E5

Visby, *Sw.* **91** D3

Viscount Melville Sound, *Nunavut, Can.* **40** A3

Vishakhapatnam, *India* **117** D4

Vistula, *river, Pol.* **79** C4

Viterbo, *It.* **92** C1

Viti Levu, *island, Fiji* **160** C3

Vitim, *Russ.* **95** D4

Vitória, *Braz.* **71** D6

Vitória da Conquista, *Braz.* **71** C6

Vitsyebsk, *Belarus* **91** E5

Vizianagaram, *India* **117** D4

Vladimir, *Russ.* **77** B5

Vladivostok, *Russ.* **95** D6

Volcano Islands, *Japan* **101** C6

Volga, *river, Russ.* **79** C5

Volga River Delta, *Russ.* **114** C1

Volgograd (Stalingrad), *Russ.* **94** D1

Volgograd Reservoir, *Russ.* **79** C5

Volgogradskoye Vodokhranilishche, *Russ.* **77** C5

Volonga, *Russ.* **77** A5

Vólos, *Gr.* **93** D4

Vol'sk, *Russ.* **114** A1

Volta, Lake, *Ghana* **138** E3

Voltaire, Cape, *W. Austral., Austral.* **150** A2

Vopnafjördur, *Ice.* **76** A2

Vorkuta, *Russ.* **94** C3

Voronezh, *Russ.* **94** C1

Vostok, *research station, Antarctica* **167** C5

Vostok, Lake, *Antarctica* **167** C5

Voyageurs National Park, *U.S.* **47** A4

Voznesens'k, *Ukr.* **93** A5

Vrangelya, Ostrov (Wrangel Island), *Russ.* **95** A5

Vryburg, *S. Af.* **144** D3

Vryheid, *S. Af.* **144** D3

Vyazemskiy, *Russ.* **95** D6

Vyaz'ma, *Russ.* **91** E5

Vyborg, *Russ.* **90** C4

Vyshniy Volochek, *Russ.* **91** D5

W

Wabē Gestro, *river, Eth.* **141** E4

Waco, *Tex., U.S.* **42** D3

Wad Medani, *Sudan* **141** D3

Wagga Wagga, *N.S.W., Austral.* **149** E5

Wahiawā, *Hawai'i, U.S.* **45** D3

Waialua Bay, *Hawai'i, U.S.* **45** D2

Waigeo, *island, Indonesia* **123** D5

Waikīkī Beach, *Hawai'i, U.S.* **45** E3

Wailuku, *Hawai'i, U.S.* **45** E4

Waimea, *Hawai'i, U.S.* **45** D1

Waimea (Kamuela), *Hawai'i, U.S.* **45** E5

Waingapu, *Indonesia* **123** E4

Wainwright, *Alas., U.S.* **44** A3

Waipahu, *Hawai'i, U.S.* **45** D3

Waipi'o Bay, *Hawai'i, U.S.* **45** E5

Wajir, *Kenya* **143** A6

Wakasa Wan, *Japan* **121** D5

Wakayama, *Japan* **121** D5

Wakkanai, *Japan* **121** A6

Wales, *region, U.K.* **88** B3

Walgett, *N.S.W., Austral.* **149** D5

Wallaroo, *S. Austral., Austral.* **149** E4

Walla Walla, *Wash., U.S.* **42** A1

Wallis, Îles, *Fr.* **161** C4

Walvis Bay, *Namibia* **144** C1

Wamba, *river, Angola, Dem. Rep. of the Congo* **142** C2

Wami, *river, Tanzania* **143** C6

Wando, *S. Kor.* **120** E2

Waranga Basin, *Vic., Austral.* **151** E5

Warangal, *India* **117** D4

Warmbad, *Namibia* **144** D2

Warrego, *river, Qnsld., Austral.* **151** D5

Warrenton, *S. Af.* **144** D2

Warri, *Nigeria* **139** E4

Warrnambool, *Vic., Austral.* **149** E5

Warsaw *see* Warszawa, *Pol.* **91** E3

Warszawa (Warsaw), *Pol.* **91** E3

Washington, *D.C., U.S.* **43** C6

Washington, *state, U.S.* **42** A1

Waskaganish, *Que., Can.* **41** D4

Waterford (Port Láirge), *Ire.* **88** B2

Watling *see* San Salvador, *island, Bahamas* **51** B4

Watson Lake, *Yukon, Can.* **40** B1

Wau, *S. Sudan* **141** E2

Webi Jubba, *river, Somalia* **141** E4

Weddell Sea, *Antarctica* **166** B2

Weipa, *Qnsld., Austral.* **149** A5

Welkom, *S. Af.* **144** D3

Wellington, *N.Z.* **160** E3

Wellington, Isla, *Chile* **73** E1

Wellington Island, *Chile* **59** F2

X

Y

CREDITS AND SOURCES

Cover, Half-title, full-title

Image: Félix Pharand-Deschênes / Globaïa. globaia.org
Data Sources: *Paved and unpaved roads, pipelines, railways, and transmission lines*: VMap0, National Geospatial-Intelligence Agency, Sept. 2000; *Shipping lanes*: NOAA's SEAS BBXX database, from Oct. 2004 to Oct. 2005. *Air networks*: International Civil Aviation Organization statistics. *Urban areas*: naturalearthdata.com; *Submarine cables*: Greg Mahlknecht's Cable Map; *Earth texture*: VMAP0; NOAA's SEAS BBXX database; Tom Patterson/Natural Earth: naturalearthdata.com

Physical and Political Maps and Editorial Content

Bathymetric Relief: ETOPO1: 1 Arc-Minute Global Relief Model: NOAA Technical Memorandum National Environmental Satellite, Data, and Information Service (NESDIS) National Geophysical Data Center (NGDC), March 2009.
Topographic Relief: GTOPO30, United States Geological Survey (USGS)
Land cover data: Tom Patterson/Natural Earth: naturalearthdata.com

Central Intelligence Agency
Library of Congress,
  Geography and Map Division
National Aeronautics and Space Administration (NASA)
  Earth Observatory System (EOS)
National Geospatial-Intelligence Agency
  U.S. Board on Geographic Names (BGN)
Naval Research Laboratory
National Oceanic and Atmospheric Administration (NOAA)
  National Climatic Data Center

National Geophysical Data Center
National Ocean Service (NOS)
U.S. Census Bureau
U.S. Department of Defense
U.S. Department of the Interior
  Bureau of Land Management (BLM)
  Geological Survey (USGS)
  National Park Service
  Office of Territories
U.S. Department of State
U.S. Naval Oceanographic Office
U.S. Navy/NOAA Joint Ice Center
Population Reference Bureau (PRB)
United Nations (UN)

World Thematic Maps

Population, pages 16–17

Population Density map: Source: Landscan 2014™ Population Dataset created by UT-Battelle, LLC, the management and operating contractor of the Oak Ridge National Laboratory acting on behalf of the U.S. Department of Energy under Contract No. DE-AC05-00OR22725. Distributed by East View Geospatial: geospatial.com and East View Information Services: eastview.com/online/landscan
Regional Population Growth graph: © National Geographic Society.

Land Cover, pages 18–19

Global Land Cover composition: Boston University Department of Geography and Environment Global Land Cover Project. Source data provided by NASA's Moderate Resolution Imaging Spectraradiometer.

Plate Tectonics, pages 20–21

Plate Tectonics map: *Earthquake data:* USGS Earthquake Hazards Program and USGS National Earthquake Information Center (NEIC). earthquake.usgs.gov. *Volcanism data:* Smithsonian Institution, Global Volcanism Program. volcano.si.edu; USGS and the International Association of Volcanology and Chemistry of the Earth's

Interior. vulcan.wr.usgs.gov.
Tectonic block diagrams: Chuck Carter.

Climate, pages 22–23

Modified Köppen classification map: © H. J. de Blij, P. O. Muller, and John Wiley & Sons, Inc.

Geographic Extremes

World, 24-25; North America, 26-27; South America, 54-55; Europe, 74-75; Asia, 96-97; Africa, 124-125; Australia & Oceania, 146-147: **Largest Cities by Population:** *World Urbanization Prospects: The 2014 Revision,* United Nations Department of Economic and Social Affairs, Population Division. **Longest Rivers:** Geoscience Australia; Natural Resources Canada; U.S. Department of Commerce, U.S. Census Bureau, *Statistical Abstract of the United States, 2012.* **Largest Lakes:** Geoscience Australia; NASA; UNESCO. *National Geographic Atlas of the World, 10th Edition* © 2015 National Geographic Society

Continental Thematic Maps

Continental Population

North America, 32-33; South America, 60-61; Europe, 80-81; Asia, 102-103; Africa, 130-131; Australia and Oceania, 152-153: **Population Density:** Source: Landscan 2014™ Population Dataset created by UT-Battelle, LLC, the management and operating contractor of the Oak Ridge National Laboratory acting on behalf of the U.S. Department of Energy under Contract No. DE-AC05-00OR22725. **Percent Population Change 2010–2050:** PRB 2010 World Population Data Sheet.

Natural Hazards

North America, 34-35; South America, 62-63; Europe, 82-83; Asia, 104-105; Africa, 132-133; Australia & Oceania, 154-155: **Natural Hazards:** USGS Earthquake Hazard Program; Smithsonian Institution's Global Volcanism Program;

National Geophysical Data Center/World Data Center (NGDC/WDC) Historical Tsunami Database.

Land Cover

North America, 36-37; South America, 64-65; Europe, 84-85; Asia, 106-107; Africa, 134-135; Australia & Oceania, 156-157: **Global Land Cover imagery:** Boston University Department of Geography and Environment Global Land Cover Project. Source data provided by NASA's Moderate Resolution Imaging Spectraradiometer.

Climate and Water

North America, 38-39; South America, 66-67; Europe, 86-87; Asia, 108-109; Africa, 136-137; Australia & Oceania, 158-159: **Modified Köppen classification maps:** © H. J. de Blij, P. O. Muller, and John Wiley & Sons, Inc. **Water Availability maps:** Aaron Wolf, Oregon State University.

STAFF FOR THIS ATLAS

Damien Saunder, Director of Cartography

Debbie Gibbons, Director of Intracompany Cartography

Ted Sickley, Director, Cartographic Database

Greg Ugiansky, Project Manager and Map Production

Scott Zillmer, Map Editor

Rosemary Wardley, Map Editor

Irene Berman-Vaporis, Map Editor

Ryan Williams, Map Editor

NATIONAL
GEOGRAPHIC

Compact
Atlas of the World

SECOND EDITION

Since 1888, the National Geographic Society has funded more than 12,000 research, exploration, and preservation projects around the world. National Geographic Partners distributes a portion of the funds it receives from your purchase to National Geographic Society to support programs including the conservation of animals and their habitats.

National Geographic Partners
1145 17th Street NW
Washington, DC 20036-4688 USA

Become a member of National Geographic and activate your benefits today at natgeo.com/jointoday.

For information about special discounts for bulk purchases, please contact National Geographic Books Special Sales: specialsales@natgeo.com

For rights or permissions inquiries, please contact National Geographic Books Subsidiary Rights: bookrights@natgeo.com

Copyright © 2017 National Geographic Partners, LLC.
All rights reserved. Reproduction of the whole or any part of the contents without written permission from the publisher is prohibited.

NATIONAL GEOGRAPHIC and Yellow Border Design are trademarks of the National Geographic Society, used under license.

ISBN: 978-1-4262-1787-6

Printed in Hong Kong
17/THK/1